ANOVA and Mixed Models

ANOVA and Mixed Models: A Short Introduction Using R provides both the practitioner and researcher a compact introduction to the analysis of data from the most popular experimental designs. Based on knowledge from an introductory course on probability and statistics, the theoretical foundations of the most important models are introduced. The focus is on an intuitive understanding of the theory, common pitfalls in practice, and the application of the methods in R. From data visualization and model fitting, up to the interpretation of the corresponding output, the whole workflow is presented using R. The book does not only cover standard ANOVA models, but also models for more advanced designs and mixed models, which are common in many practical applications.

Features

- Accessible to readers with a basic background in probability and statistics
- Covers fundamental concepts of experimental design and cause-effect relationships
- Introduces classical ANOVA models, including contrasts and multiple testing
- Provides an example-based introduction to mixed models
- Features basic concepts of split-plot and incomplete block designs
- R code available for all steps
- Supplementary website with additional resources and updates available at https://stat.ethz.ch/~meier/teaching/book-anova/

This book is primarily aimed at students, researchers, and practitioners from all areas who wish to analyze corresponding data with R. Readers will learn a broad array of models hand-in-hand with R, including the applications of some of the most important add-on packages.

Lukas Meier is a senior scientist at the Seminar für Statistik at ETH Zürich. His main interests are teaching statistics at various levels, the application of statistics in many fields of applications using advanced ANOVA or regression models, and high-dimensional statistics. He co-leads the statistical consulting service at ETH Zürich and is the director of a continuing education program in applied statistics.

Chapman & Hall/CRC
The R Series

Series Editors
John M. Chambers, Department of Statistics, Stanford University, California, USA
Torsten Hothorn, Division of Biostatistics, University of Zurich, Switzerland
Duncan Temple Lang, Department of Statistics, University of California, Davis, USA
Hadley Wickham, RStudio, Boston, Massachusetts, USA

Recently Published Titles

For more information about this series, please visit: https://www.crcpress.com/
Chapman--HallCRC-The-R-Series/book-series/CRCTHERSER

ANOVA and Mixed Models

A Short Introduction Using R

Lukas Meier
ETH Zürich, Switzerland

CRC Press
Taylor & Francis Group
Boca Raton London New York

CRC Press is an imprint of the
Taylor & Francis Group, an **Informa** business

A CHAPMAN & HALL BOOK

First edition published 2023
by CRC Press
6000 Broken Sound Parkway NW, Suite 300, Boca Raton, FL 33487-2742

and by CRC Press
4 Park Square, Milton Park, Abingdon, Oxon, OX14 4RN

Library of Congress Cataloging-in-Publication Data

Names: Meier, Lukas, author.
Title: ANOVA and mixed models : a short introduction using R / Lukas Meier, ETH Zürich, Switzerland.
Description: First edition. | Boca Raton : Chapman & Hall/CRC Press, 2023.
| Series: Chapman and Hall/CRC the R series | Includes bibliographical references and index.
Identifiers: LCCN 2022022112 (print) | LCCN 2022022113 (ebook) | ISBN 9780367704223 (hardback) | ISBN 9780367704209 (paperback) | ISBN 9781003146216 (ebook)
Subjects: LCSH: Analysis of variance. | R (Computer program language)
Classification: LCC QA279 .M3886 2023 (print) | LCC QA279 (ebook) | DDC 519.5/3802855133--dc23/eng20220910
LC record available at https://lccn.loc.gov/2022022112
LC ebook record available at https://lccn.loc.gov/2022022113

ISBN: 978-0-367-70422-3 (hbk)
ISBN: 978-0-367-70420-9 (pbk)
ISBN: 978-1-003-14621-6 (ebk)

DOI: 10.1201/ 9781003146216

Typeset in Latin modern
by KnowledgeWorks Global Ltd.

Publisher's note: This book has been prepared from camera-ready copy provided by the authors.

For Jeannine, Andrin, Silvan and Dario

Contents

Preface

This book should help you get familiar with analysis of variance (ANOVA) and mixed models in R (R Core Team, 2021). From a methodological point of view, we build upon the knowledge of an introductory course to probability and statistics covering the basic concepts of statistical inference (estimation, hypothesis tests, confidence intervals) up to the two-sample t-test. See for example Dalgaard (2008) for an introduction of both theory and the corresponding functions in R. A more theoretical reference is Rice (2007).

There are of course already well-established excellent textbooks covering ANOVA including experimental design in great detail. Examples are Oehlert (2000), Kuehl (2000), Montgomery (2019) and many more. We build upon these great books. From a mathematical point of view, we use similar notation as Oehlert (2000). The goal of this book is to provide a *compact* overview of the most important topics including the corresponding applications in R using flexible mixed model approaches. We also use examples from the classical textbooks and will redo the corresponding statistical analyses in R.

As this is an introductory text, the focus is on getting to know multiple experimental design types, when they are being used and what a proper analysis in R looks like. This is why we will not do all the details, especially for the more advanced topics. The idea is that if the reader is familiar with the basic concepts and their applications in R, this knowledge can be extended (and applied) to other areas.

Besides discussing the theory and the corresponding R functions, we also try to give you an intuition in when and how things can go wrong and what aspects have to be considered in practice. This is not only useful when planning an experiment on your own, but also when analyzing data from other sources or when reading a research paper.

From a statistical point of view, an ANOVA model is nothing more than a special case of a linear regression model. Note that no prior knowledge of linear regression is needed for this book. For the basic models, we mostly use the function aov in R in order to get the "classical" outputs. In fact, aov simply calls lm (the linear regression model fitting function) and adjusts the output accordingly. We sometimes mention extensions to more general linear regression models. However, this book is not meant to be an introductory text to linear regression. See for example Fox and Weisberg (2019) or Faraway (2005) for applied introductions.

If not stated otherwise, we use a significance level of 5% if we make statements about statistical significance, or equivalently, a coverage level of 95% for the corresponding confidence intervals.

If you find any errors, inconsistencies or if you miss something, please e-mail[1] me or fill out the anonymous feedback form at https://goo.gl/ZBvjj9.

The most recent version of this book and a list of errors can be found on https://stat.ethz.ch/~meier/teaching/book-anova/.

Structure of the Book

We begin with a non-technical introduction to the general principles of experimental design in Chapter 1. Chapter 2 then introduces the first models for designs with only one factor. More specific questions regarding these models are then discussed in Chapter 3, including the problem of multiple testing. Chapter 4 introduces factorial designs which arise if a treatment is a combination of multiple factors. A short introduction to complete block designs, which are a great way to increase power or precision, can be found

---

[1] meier@stat.math.ethz.ch

in Chapter 5. Chapter 6 introduces a new class of models including random and fixed effects, the so-called mixed models which are very popular in many applied areas. Some more special designs follow: Chapter 7 introduces a new class of designs which can deal with experimental units of different sizes, the so-called split-plot designs. We conclude with Chapter 8 about block designs with small blocks that cannot accommodate all treatments, so-called incomplete block designs.

Software Information and Conventions

This book uses a lot of R code. If you are completely new to R, you can get more information for example at https://cran.r-project.org/manuals.html or https://education.rstudio.com/.

The R code and output has the following form:

```
text <- "Let's get started ..."
paste(text, "now!", sep = " ")
```

```
## [1] "Let's get started ... now!"
```

This means that output lines start with two comments sign "##". For better readability, we sometimes shorten the R output a bit. If we remove multiple lines, this will be indicated with the symbol "## ...", i.e., two comment signs and three dots, in the output.

Regarding plots, we mostly use base R graphics. For more complex plots we switch to ggplot2 (Wickham, 2016).

We often load data directly from the web, either in tabular format using the function read.table, or already as an R object, using the function readRDS.

The packages knitr (Xie, 2015) and bookdown (Xie, 2021) were used to compile this book.

Acknowledgments

First, I'd like to thank all members of the Seminar für Statistik at ETH Zürich for such a nice and fun working and research environment and for making it possible to work on this project. I learned a lot a long time ago from a wise man nicknamed "Puma" while working in a building named "LEO". Hans-Rudolf Roth, you are missed!

Many people contributed in various ways to this book, special thanks go to Peter Bühlmann, Markus Kalisch, Marloes Maathuis, Christoph Buck, Claude Renaux, Camilla Gerboth, Tanja Finger, Michael Zellinger, Reto Zihlmann and Bill Perry.

I also want to thank Rob Calver from Chapman & Hall/CRC Press for the support and patience.

Finally, and most importantly, I would like to thank my family for all the support.

Lukas Meier
Zürich, Switzerland

1

Learning from Data

1.1 Cause-Effect Relationships

Why do people like to collect (nowadays, big) data? Typically, the goal is to find "relationships" between different "variables": What fertilizer combination makes my plants grow tall? What medication reduces headache most? Is a new vaccine effective enough?

From a more abstract point of view, we are in the situation where we have a "system" or a "process" (e.g., a plant) with many input variables (so-called predictors) and an output (so-called response), see also Montgomery (2019). In the previous example the ingredients of the fertilizer are the inputs, and the output could be the biomass of a plant.

Ideally, we would like to find **cause-effect relationships**, meaning that when we actively change one of the inputs, i.e., we make an **intervention** on the system, this will cause the output to change. If we can just observe a system under different settings (so-called **observational studies** or **observational data**), it is much harder to make a statement about causal effects, as can be seen in the following examples:

- Is the seatbelt sign on an airplane causing a plane to shake? If we could switch it on ourselves, would the plane start shaking? See *The Family Circus* comic strip (Keane, 1998): "I wish they didn't turn on that seatbelt sign so much! Every time they do, it gets bumpy."

- Are ice cream sales causing people to drown? If we would stop selling ice cream, would drowning decrease, or even stop?

With observational data, we can typically just make a statement about an **association** between two variables. One potential danger is the existence of **confounding variables** or **confounders**. A confounder is a common cause for two variables. In the previous examples we had the following situations:

- Turbulent weather simultaneously makes the pilot switch on the seatbelt sign and the plane shake. What we *observe* is an *association* between the appearance of the seatbelt sign and a shaking plane. The seatbelt sign is *not* a cause of the shaking plane.

- Hot weather makes people want to go swimming and at the same time is beneficial for ice cream sales. What we *observe* is an *association* between ice cream sales and the number of drowning incidents. Ice cream sales is *not* a cause of the number of drowning incidents.

A more classical example was the question whether smoking causes lung cancer or maybe "bad genes" make people smoke and develop lung cancer at the same time (Stolley, 1991). More examples of spurious associations can be found on http://www.tylervigen.com/spurious-correlations.

To find out cause-effect relationships, we should ideally be able to make an intervention on a system ourselves. If we would occasionally switch on the seatbelt sign, ignoring the weather conditions, we would see that the plane will typically not start shaking. This is what we do in **experimental studies**. We *actively change* the inputs of the system, i.e., we make an intervention, and we observe what happens with the output. It is like reverse engineering the system to find out how it works. This is also how little kids or babies discover the world: "If I push this button then ...".

Remark: It is also possible to make a statement about causal effects using observational data. To do so, we would need to know the underlying "causal diagram", typically unknown in practice, where direct causal effects are visualized by arrows. A set of rules (see for example Pearl and Mackenzie, 2018) would then tell us what variables we have to consider in our analysis ("conditioning"). In the seatbelt sign example, if we also consider weather conditions, we would see that there is no causal effect from the seatbelt sign on the movements of the plane. The corresponding causal diagram would look like the one in Figure 1.1.

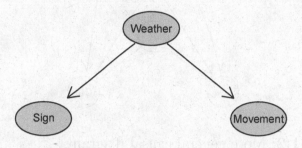

FIGURE 1.1: Causal diagram for the seatbelt sign example.

A more complicated example is illustrated in Figure 1.2. If we are interested in the total causal effect of "In" on "Out" using observational data, we would need to believe that this causal diagram is really representing the truth (in particular, we did not forget any important variables) and we have to derive the correct set of variables to condition on (here, only "D"). On the other hand, if we can do an experiment, we simply make an intervention on variable "In" and see what happens with the output "Out," we do not have to know the underlying causal diagram.

1.2 Experimental Studies

In this section, we closely follow Chapter 1 of Oehlert (2000) to give you an idea of the main ingredients of an experimental study.

Before designing an experimental study, we must have a **focused** and **precise research question** that we want to answer with experimental data, e.g., "how does fertilizer A compare to fertilizer B with respect to biomass of plants after 10 weeks?". Quite often, people collect large amounts of data and afterward think, "let's see

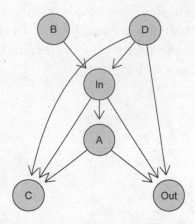

FIGURE 1.2: More general causal diagram.

whether we find some interesting patterns in our data!". Such an approach is permissible in order to **create** some research question. However, we focus on the part where we want to **confirm** a certain specific conjecture. We have to make sure that it is actually testable, i.e., that we can do the appropriate interventions and that we can measure the right response (see Section 1.2.4).

An experimental study consists of the following ingredients:

- The different interventions which we perform on the system: the different **treatments**, e.g., different fertilizer combinations that we are interested in, but also other predictors (inputs) of the system (see Section 1.2.1).

- **Experimental units**: the actual objects on which we apply the treatments, e.g., plots of land, patients, fish tanks, etc. (see Section 1.2.3).

- A method that assigns experimental units to treatments, typically **randomization** or restricted versions thereof (see Section 1.2.2).

- **Response(s)**: the output that we measure, e.g., biomass of plants (see Section 1.2.4).

In addition, when designing an experimental study, the **analysis** of the resulting data should already be considered. For example, we need an idea of the **experimental error** (see Section 1.2.5); otherwise, we *cannot* do any statistical inference. It is always a good idea to try to analyze some simulated data before performing the experiment, as this can potentially already reveal some serious flaws of the design.

1.2.1   Predictors or Treatments

We distinguish between the following types of predictors (inputs of the system):

- Predictors that are of primary interest and that can (ideally) be varied according to our "wishes": the conditions we want to compare, or the **treatments**, e.g., fertilizer type.

- Predictors that are systematically recorded such that potential effects can later be eliminated in our calculations ("controlling for ..."), e.g., weather conditions. In this context, we also call them **covariates**.

- Predictors that can be kept constant and whose effects are therefore eliminated, e.g., always using the same measurement device.

- Predictors that we can neither record nor keep constant, e.g., some special soil properties that we cannot measure.

1.2.2   Randomization

Randomization, i.e., the *random* allocation of experimental units to the different treatments, ensures that the only systematic difference between the different treatment groups is the treatment: No matter what special properties the experimental units have, we have a similar mix in each group. This protects us from confounders and is the reason why a properly randomized experiment allows us to make a statement about a cause-effect relationship between treatment and response.

The importance of randomization is also nicely summarized in the following quote:

"Randomization generally costs little in time and trouble, but it can save us from disaster." (Oehlert, 2000)

We can and should also randomize other aspects like the sequence in which experimental units are measured. This protects us from time being a confounder.

Quite often, we already know that some experimental units are more alike than others *before* doing the experiment. Think for example of different locations for an agricultural experiment. Typically, we then do a randomization "within" homogeneous blocks (here, at each location). This restricted version of randomization is called **blocking**. A **block** is a subset of experimental units that is more homogeneous than the entire set. Blocking typically increases *precision* of an experiment, see Chapter 5. The rule to remember is:

"Block what you can; randomize what you cannot." (Box et al., 1978)

1.2.3   Experimental and Measurement Units

An **experimental unit** is defined as the object on which we apply the treatments by randomization. The general rule is (Oehlert, 2000): "An experimental unit should be able to receive any treatment *independently* of the other units". On the other hand, a **measurement unit** is the object on which the response is being *measured*. There are many situations where experimental units and measurement units are *not* the same. This can cause severe consequences on the analysis if not treated appropriately.

For example, if we randomize different food supplies to fish tanks (containing multiple fish), the experimental unit is given by a fish tank and *not* an individual fish. However, a measurement unit will be given by an individual fish, e.g., we could take as response the body weight of a fish after 5 weeks. Typically, we *aggregate* the values of the measurement units such that we get *one* value per experimental unit, e.g., take the average body weight per fish tank. These values will typically be the basis for the statistical analysis later on.

There are also situations where there are different "sizes" (large and small) of experimental units within the same experiment, see Chapter 7.

As we want our results to have broad validity, the experimental units should ideally be a *random* sample from the population of interest. If the experimental units do not represent the population well, drawing conclusions from the experimental results to the population will be challenging. In this regard, **internal validity** means that the study was set up and performed correctly with the experimental units *at hand*, e.g., by using randomization, blinding as mentioned below, etc. Hence, the conclusions that we draw are valid for the "study population". On the other hand, **external validity** means that the conclusions can be transferred to a more general population. For example, if we properly run a psychological study with volunteers, we have internal validity. However, we do not necessarily have external validity if the volunteers do not represent the general population well (because they are not a random sample).

1.2.4 Response

The **response**, or the output of the process, should be chosen such that it reflects useful information about the process under study, e.g., body weight after some diet treatment or biomass when comparing different fertilizers. It is *your* responsibility that the response is a reasonable quantity to study your research hypothesis.

There are situations where the response is *not* directly measurable. An example is HIV progression. In cases like this, we need a so-called **surrogate response**. For HIV, this could be CD4-cell counts. These cells are being attacked by HIV and the number gives you an idea of the health status of the immune system.

1.2.5 Experimental Error

Consider the following hypothetical example: We make an experiment using two plants. One gets fertilizer A and the other one B. After 4 weeks, we observe that the plant receiving fertilizer A has twice the biomass of the plant receiving fertilizer B. Can we conclude that fertilizer A is causing larger biomass? Unfortunately, we cannot say so in this situation, even if we randomized the two fertilizers to the two plants. The experiment does not give us any information whether the difference that we observe is larger than the natural variation from plant to plant getting the same fertilizer. It could very well be the case that there is no difference between the two fertilizers, meaning that the difference that we observe is just natural variation from plant to plant. From a technical point of view, this is like trying to do a two-sample t-test with only one observation in each group.

In other words, no two experimental units are perfectly identical. Hence, it is completely natural that we will measure slightly different values for the response, even if they get the same treatment. We should design our experiment such that we get an idea of this so-called **experimental error**. Here, this means we would need multiple plants (replicates) receiving the same treatment. If the difference between the treatments is much larger than the experimental error, we can conclude that there is a treatment effect caused by the fertilizer. You've learned analyzing such data with a two-sample t-test, which would not work with only one observation in each treatment group, see above.

A "true" replicate should be given by another independent experimental unit. If we would measure the biomass of the original two plants 10 times each, we would also have multiple measurements per treatment group. However, the error that we observe is simply the measurement error of our measurement device. Hence, we still have no clue about the experimental error. We would call them **pseudoreplicates**. From a technical point of view, we could do a two-sample *t*-test in this situation. However, we could just conclude that these two specific plants are different for whatever reason. This is *not* what you typically want to do.

1.2.6 More Terminology

Blinding means that those who measure or assess the response do not know which treatment is given to which experimental unit. With humans it is common to use **double-blinding** where in addition the patients do not know the assignment either. Blinding protects us from (unintentional) bias due to expectations.

A **control treatment** is typically a standard treatment with which we want to compare. It can also be no treatment at all. You should always ask yourself, "How does it compare to no treatment or the standard treatment?". For example, a physiotherapist claims that a new therapy after surgery reduces pain by about 30%. However, patients *without* any therapy after surgery might have a reduction of 50%!

1.2.7 A Few Examples

What's wrong with the following examples?

- Mike is interested in the difference between two teaching techniques. He randomly selects 10 Harvard lecturers that apply technique *A* and 10 MIT lecturers that apply technique *B*. Each lecturer reports the average grade of his class.

- Linda has two cages of mice. Mice in cage 1 get a special food supply, while mice in cage 2 get ordinary food (control treatment).

- Melanie offers a new exam preparation course. She claims that, on average, only 20% of her students fail the exam.

2

Completely Randomized Designs

We assume for the moment that the experimental units are *homogeneous*, i.e., no restricted randomization scheme is needed (see Section 1.2.2). For example, this is a reasonable assumption if we have 20 similar plots of land (experimental units) at a single location. However, the assumption would not be valid if we had five different locations with four plots each.

We will start with assigning experimental units to treatments and then do a proper statistical analysis. In your introductory course to statistics, you learned how to compare two independent groups using the two-sample t-test. If we have more than two groups, the t-test is *not* directly applicable anymore. Therefore, we will develop an extension of the two-sample t-test for situations with more than two groups. The two-sample t-test will still prove to be very useful later if we do pairwise comparisons between treatments, see Section 3.2.4.

2.1 One-Way Analysis of Variance

On an abstract level, our goal consists of comparing $g \geq 2$ treatments, e.g., $g = 4$ different fertilizer types. As available resources, we have N experimental units, e.g., $N = 20$ plots of land, that we assign *randomly* to the g different treatment groups having n_i observations each, i.e., we have $n_1 + \cdots + n_g = N$. This is a so-called **completely randomized design (CRD)**. We simply randomize the experimental units to the different treatments and are *not* considering any other structure or information, like location, soil properties, etc. This is the most elementary experimental design and basically the building block of all more complex designs later. The optimal choice (with respect to power) of $n_1, ..., n_g$ depends on our research question. If all the treatment groups have the same number of experimental units, we call the design **balanced**; this is usually a good choice unless there is a special control group with which we want to do a lot of comparisons.

If we want to randomize a total of 20 experimental units to the four different treatments labelled A, B, C and D using a balanced design with five experimental units per treatment, we can use the following R code.

```
treat.ord <- rep(c("A", "B", "C", "D"), each = 5)
## could also use LETTERS[1:4] instead of c("A", "B", "C", "D")
treat.ord
```

```
##  [1] "A" "A" "A" "A" "A" "B" "B" "B" "B" "B" "C" "C" "C" "C"
## [15] "C" "D" "D" "D" "D" "D"
```

```
sample(treat.ord) ## random permutation
```

```
##  [1] "B" "D" "C" "B" "B" "B" "A" "A" "C" "C" "D" "A" "B" "C"
## [15] "D" "C" "D" "D" "A" "A"
```

This means that the first experimental unit will get treatment B, the second D, and so on.

2.1.1   Cell Means Model

In order to do statistical inference, we start by formulating a parametric model for our data. Let y_{ij} be the observed value of the response of the jth experimental unit in treatment group i, where $i = 1, ..., g$ and $j = 1, ..., n_i$. For example, y_{23} could be the biomass of the third plot getting the second fertilizer type.

In the so-called **cell means model**, we allow each treatment group, "cell", to have its *own* expected value, e.g., expected biomass, and we assume that observations are independent (also across different treatment groups) and fluctuate around the corresponding expected value according to a normal distribution. This means the observed value y_{ij} is the realized value of the random variable

$$Y_{ij} \sim N(\mu_i, \sigma^2), \text{ independent} \qquad (2.1)$$

where

- μ_i is the expected value of treatment group i,
- σ^2 is the variance of the normal distribution.

As for the standard two-sample t-test, the variance is assumed to be *equal* for all groups. We can rewrite Equation (2.1) as

$$Y_{ij} = \mu_i + \epsilon_{ij} \qquad (2.2)$$

with random errors ϵ_{ij} i.i.d. $\sim N(0, \sigma^2)$ (i.i.d. stands for independent and identically distributed). We simply separated the normal distribution around μ_i into a deterministic part μ_i and a stochastic part ϵ_{ij} fluctuating around *zero*. This is nothing other than the experimental error. As before, we say that Y is the response and the treatment allocation is a categorical predictor. A categorical predictor is also called a **factor**. We sometimes distinguish between **unordered (or nominal)** and **ordered (or ordinal) factors**. An example of an unordered factor would be fertilizer type, e.g., with levels "A", "B", "C" and "D", and an example of an ordered factor would be income class, e.g., with levels "low", "middle" and "high".

For those who are already familiar with linear regression models, what we have in Equation (2.2) is nothing more than a regression model with a single categorical predictor and normally distributed errors.

We can rewrite Equation (2.2) using

$$\mu_i = \mu + \alpha_i \qquad (2.3)$$

for $i = 1, \ldots, g$ to obtain

$$Y_{ij} = \mu + \alpha_i + \epsilon_{ij} \qquad (2.4)$$

with ϵ_{ij} i.i.d. $\sim N(0, \sigma^2)$.

We also call α_i the ith **treatment effect**. Think of μ as a "global mean" and α_i as a "deviation from the global mean due to the ith treatment." We will soon see that this interpretation is not always correct, but it is still a helpful way of thinking. At first sight, this looks like writing down the problem in a more complex form. However, the formulation in Equation (2.4) will be very useful later if we have more than one treatment factor and want to "untangle" the influence of multiple treatment factors on the response, see Chapter 4.

If we carefully inspect the parameters of models (2.2) and (2.4) we observe that in (2.2) we have the parameters $\mu_1, \dots, \mu_g$ and σ^2, while in (2.4) we have $\mu, \alpha_1, \dots, \alpha_g$ and σ^2. We (silently) introduced one additional parameter. In fact, model (2.4) is not identifiable anymore because we have $g + 1$ parameters to model the g mean values μ_i. Or in other words, we can "shift around" effects between μ and the α_i's without changing the resulting values of μ_i, e.g., we can adjust $\mu + 10$ and $\alpha_i - 10$ leading to the same μ_i's. Hence, we need a side constraint on the α_i's that "removes" that additional parameter. Typical choices for such a constraint are given in Table 2.1 where we have also listed the interpretation of the parameter μ and the corresponding naming convention in R. For ordered factors, other options might be more suitable, see Section 2.6.1.

TABLE 2.1: Different side constraints.

Name	Side Constraint	Meaning of μ	R
weighted sum-to-zero	$\sum_{i=1}^{g} n_i \alpha_i = 0$	$\mu = \frac{1}{N} \sum_{i=1}^{g} n_i \mu_i$	-
sum-to-zero	$\sum_{i=1}^{g} \alpha_i = 0$	$\mu = \frac{1}{g} \sum_{i=1}^{g} \mu_i$	`contr.sum`
reference group	$\alpha_1 = 0$	$\mu = \mu_1$	`contr.treatment`

For all of the choices in Table 2.1 it holds that μ determines some sort of "global level" of the data and α_i contains information about differences between the group means μ_i from that "global level."

At the end, we arrive at the very same $\mu_i = \mu + \alpha_i$ for all choices. This means that the destination of our journey (μ_i) is always the same, but the route we take (μ, α_i) is different.

Only $g-1$ elements of the treatment effects are allowed to vary freely. In other words, if we know $g-1$ of the α_i values, we automatically know the remaining α_i. We also say that the treatment effect has $g-1$ **degrees of freedom (df)**.

In R, the side constraint is set using the option `contrasts` (see examples below). The default value is `contr.treatment` which is the side constraint "reference group" in Table 2.1.

2.1.2 Parameter Estimation

We estimate the parameters using the **least squares criterion** which ensures that the model fits the data well in the sense that the squared deviations from the observed data y_{ij} to the model values $\mu_i = \mu + \alpha_i$ are minimized, i.e.,

$$\hat{\mu}, \hat{\alpha}_i = \text{argmin}_{\mu, \alpha_i} \sum_{i=1}^{g} \sum_{j=1}^{n_i} \left(y_{ij} - \mu - \alpha_i\right)^2.$$

Or equivalently, when working directly with the μ_i's, we get

$$\hat{\mu}_i = \text{argmin}_{\mu_i} \sum_{i=1}^{g} \sum_{j=1}^{n_i} \left(y_{ij} - \mu_i\right)^2.$$

We use the following notation:

$$y_{i\cdot} = \sum_{j=1}^{n_i} y_{ij} \qquad\qquad \text{sum of group } i$$

$$y_{\cdot\cdot} = \sum_{i=1}^{g} \sum_{j=1}^{n_i} y_{ij} \qquad\qquad \text{sum of all observations}$$

$$\overline{y}_{i\cdot} = \frac{1}{n_i} \sum_{j=1}^{n_i} y_{ij} \qquad\qquad \text{mean of group } i$$

$$\overline{y}_{..} = \frac{1}{N} \sum_{i=1}^{g} \sum_{j=1}^{n_i} y_{ij} \qquad \text{overall (or total) mean}$$

As we can independently estimate the values μ_i of the different groups, one can show that $\hat{\mu}_i = \overline{y}_{i.}$, or in words, the estimate of the expected value of the response of the ith treatment group is the mean of the corresponding observations. Because of $\mu_i = \mu + \alpha_i$ we have

$$\hat{\alpha}_i = \hat{\mu}_i - \hat{\mu}$$

which together with the meaning of μ in Table 2.1 we can use to get the parameter estimates $\hat{\alpha}_i$'s. For example, using the weighted sum-to-zero constraint we have $\hat{\mu} = \overline{y}_{..}$ and $\hat{\alpha}_i = \hat{\mu}_i - \hat{\mu} = \overline{y}_{i.} - \overline{y}_{..}$. On the other hand, if we use the reference group side constraint we have $\hat{\mu} = \overline{y}_1$ and $\hat{\alpha}_i = \hat{\mu}_i - \hat{\mu} = \overline{y}_{i.} - \overline{y}_1$ for $i = 2, ..., g$.

Remark: The values we get for the $\hat{\alpha}_i$'s (heavily) depend on the side constraint that we use. In contrast, this is *not* the case for $\hat{\mu}_i$.

The estimate of the error variance $\hat{\sigma}^2$ is also called **mean squared error** MS_E. It is given by

$$\hat{\sigma}^2 = MS_E = \frac{1}{N-g} SS_E,$$

where SS_E is the **error** or (**residual**) **sum of squares**

$$SS_E = \sum_{i=1}^{g} \sum_{j=1}^{n_i} (y_{ij} - \hat{\mu}_i)^2.$$

Remark: The deviation of the observed value y_{ij} from the estimated cell mean $\hat{\mu}_i$ is known as the **residual** r_{ij}, i.e.,

$$r_{ij} = y_{ij} - \hat{\mu}_i.$$

It is an "estimate" of the error ϵ_{ij}, see Section 2.2.1.

We can also write

$$MS_E = \frac{1}{N-g} \sum_{i=1}^{g} (n_i - 1) s_i^2,$$

where s_i^2 is the empirical variance in treatment group i, i.e.,

$$s_i^2 = \frac{1}{n_i - 1} \sum_{j=1}^{n_i} (y_{ij} - \hat{\mu}_i)^2.$$

As in the two-sample situation, the denominator $N - g$ ensures that $\hat{\sigma}^2$ is an unbiased estimator for σ^2. We also say that the error estimate has $N - g$ degrees of freedom. A rule to remember is that the error degrees of freedom are given by the total number of observations (N) minus the number of groups (g).

Let us have a look at an example using the built-in data set `PlantGrowth` which contains the dried weight of plants under a control and two different treatment conditions with 10 observations in each group (the original source is Dobson, 1983). See `?PlantGrowth` for more details.

```
data(PlantGrowth)
str(PlantGrowth)
```

```
## 'data.frame':    30 obs. of  2 variables:
##  $ weight: num  4.17 5.58 5.18 6.11 4.5 4.61 ...
##  $ group : Factor w/ 3 levels "ctrl","trt1",..: 1 1 1 1 1 ..
```

As we can see in the R output, `group` is a factor (a categorical predictor) having three levels, the reference level, which is the first level, is `ctrl`. The corresponding treatment effect will be set to zero when using the side constraint "reference group" in Table 2.1. We can also get the levels using the function `levels`.

```
levels(PlantGrowth[,"group"])
```

```
## [1] "ctrl" "trt1" "trt2"
```

If we wanted to change the reference level, we could do this by using the function `relevel`.

We can visualize the data by plotting `weight` vs. `group` (a so-called strip chart) or by using boxplots per level of `group`.

```
stripchart(weight ~ group, vertical = TRUE, pch = 1,
           data = PlantGrowth)
```

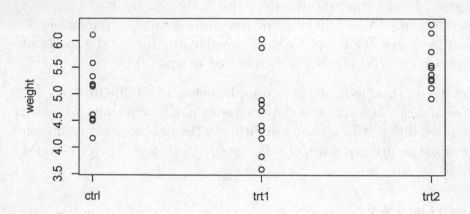

```
boxplot(weight ~ group, data = PlantGrowth)
```

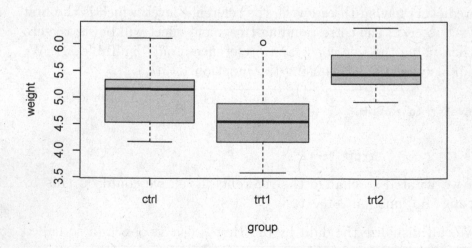

We now fit the model (2.4) using the function aov (which stands for analysis of variance). We state the model using the formula notation where the response is listed on the left-hand side and the only predictor is on the right-hand side of the tilde sign "~". The estimated parameters can be extracted using the function coef.

```
fit.plant <- aov(weight ~ group, data = PlantGrowth)
## Have a look at the estimated coefficients
coef(fit.plant)
```

```
## (Intercept)   grouptrt1   grouptrt2
##       5.032      -0.371       0.494
```

The element labelled (Intercept) contains $\hat{\mu} = 5.032$ which is the estimated expected value of the response of the *reference group* ctrl, because, by default, we use contr.treatment and the first level is the reference group; you can check the settings with the command options("contrasts"). The next element grouptrt1 is $\hat{\alpha}_2 = -0.371$. This means that the (expected) difference of group trt1 to group ctrl is estimated to be -0.371. The last element grouptrt2 is $\hat{\alpha}_3 = 0.494$. It is the difference of group trt2 to the group ctrl. Hence, for all levels except the reference level, we see differences to the reference group, while the estimate of the reference level can be found under (Intercept).

We get a clearer picture by using the function dummy.coef which lists the "full coefficients".

```
dummy.coef(fit.plant)
```

```
## Full coefficients are
##
## (Intercept):     5.032
## group:           ctrl   trt1   trt2
##                  0.000 -0.371  0.494
```

Interpretation is as in parametrization (2.3). (Intercept) corresponds to $\hat{\mu}$ and $\hat{\alpha}_1, \hat{\alpha}_2$ and $\hat{\alpha}_3$ can be found under ctrl, trt1 and trt2, respectively. For example, for $\hat{\mu}_2$ we have $\hat{\mu}_2 = 5.032 - 0.371 = 4.661$.

The **estimated cell means** $\hat{\mu}_i$, which we also call the predicted values per treatment group or simply the fitted values, can also be obtained using the function predict on the object fit.plant (which contains all information about the estimated model).

```
predict(fit.plant,
          newdata = data.frame(group = c("ctrl", "trt1", "trt2")))
```

```
##     1     2     3
## 5.032 4.661 5.526
```

Alternatively, we can use the package emmeans (Lenth, 2020), which stands for estimated marginal means. In the argument specs we specify that we want the estimated expected value for each level of group.

```
library(emmeans)
emmeans(fit.plant, specs = ~ group)
```

```
## group emmean    SE df lower.CL upper.CL
## ctrl    5.03 0.197 27     4.63     5.44
## trt1    4.66 0.197 27     4.26     5.07
## trt2    5.53 0.197 27     5.12     5.93
##
## Confidence level used: 0.95
```

Besides the estimated cell means in column emmean, we also get the corresponding 95% confidence intervals defined through columns lower.CL and upper.CL.

What happens if we change the side constraint to sum-to-zero? In R, this is called contr.sum. We use the function options to change the encoding on a global level, i.e., all subsequently fitted models will be affected by the new encoding. Note that the corresponding argument contrasts takes two values: The first one is for unordered factors and the second one is for ordered factors (we leave the second argument at its default value contr.poly for the moment).

```
options(contrasts = c("contr.sum", "contr.poly"))
fit.plant2 <- aov(weight ~ group, data = PlantGrowth)
coef(fit.plant2)
```

```
## (Intercept)      group1      group2
##       5.073      -0.041      -0.412
```

We get different values because the meaning of the parameters has changed. If we closely inspect the output, we also see that a slightly different naming scheme is being used. Instead of "factor name" and "level name" (e.g., grouptrt1) we see "factor name" and "level number" (e.g., group2). The element (Intercept) is now the global mean of the data, because the design is balanced, the sum-to-zero constraint is the same as the weighted sum-to-zero constraint, group1 is the difference of the first group (ctrl) to the global mean, group2 is the difference of the second group (trt1) to the global mean. The difference of trt2 to the global mean is not shown. Because of the side constraint we know it must be $-(-0.041 - 0.412) = 0.453$. Again, we get the full picture by calling dummy.coef.

```
dummy.coef(fit.plant2)
```

```
## Full coefficients are
##
## (Intercept):      5.073
## group:            ctrl    trt1    trt2
##                  -0.041  -0.412   0.453
```

As before, we can get the estimated cell means with predict (alternatively, we could of course again use emmeans).

```
predict(fit.plant2,
        newdata = data.frame(group = c("ctrl", "trt1", "trt2")))
```

```
##      1     2     3
## 5.032 4.661 5.526
```

Note that the output of `predict` has *not* changed. The estimated cell means do *not* depend on the side constraint that we employ. But the side constraint has a *large* impact on the meaning of the parameters $\hat{\alpha}_i$ of the model. If we do not know the side constraint, we do *not* know what the $\hat{\alpha}_i$'s actually mean!

2.1.3   Tests

With a two-sample t-test, we could test whether two samples share the same mean. We will now extend this to the $g > 2$ situation. Saying that *all* groups share the same mean is equivalent to model the data as

$$Y_{ij} = \mu + \epsilon_{ij}, \ \epsilon_{ij} \ \text{i.i.d.} \ \sim N(0, \sigma^2).$$

This is the so-called **single mean model**. It is actually a special case of the cell means model where $\mu_1 = ... = \mu_g$ which is equivalent to $\alpha_1 = ... = \alpha_g = 0$ (no matter what side constraint we use for the α_i's).

Hence, our question boils down to comparing two models, the single mean and the cell means model, which is more complex. More formally, we have the global null hypothesis

$$H_0 : \mu_1 = \mu_2 = ... = \mu_g$$

vs. the alternative

$$H_A : \mu_k \neq \mu_l \text{ for at least one pair } k \neq l.$$

We call it a *global* null hypothesis because it affects *all* parameters at the same time.

We will construct a statistical test by decomposing the variation of the response. The basic idea will be to check whether the variation between the different treatment groups (the "signal") is substantially larger than the variation within the groups (the "noise").

More precisely, total variation of the response around the overall mean can be decomposed into variation "between groups" and variation "within groups", i.e.,

$$\underbrace{\sum_{i=1}^{g}\sum_{j=1}^{n_i}(y_{ij}-\overline{y}_{..})^2}_{SS_T} = \underbrace{\sum_{i=1}^{g}\sum_{j=1}^{n_i}(\overline{y}_{i.}-\overline{y}_{..})^2}_{SS_{\text{Trt}}} + \underbrace{\sum_{i=1}^{g}\sum_{j=1}^{n_i}(y_{ij}-\overline{y}_{i.})^2}_{SS_E},$$

(2.5)

where

$$SS_T = \text{total sum of squares}$$
$$SS_{\text{Trt}} = \text{treatment sum of squares (between groups)}$$
$$SS_E = \text{error sum of squares (within groups)}$$

Hence, we have

$$SS_T = SS_{\text{Trt}} + SS_E.$$

As can be seen from the decomposition in Equation (2.5), SS_{Trt} can also be interpreted as the reduction in residual sum of squares when comparing the cell means with the single mean model; this will be a useful interpretation later.

All this information can be summarized in a so-called **ANOVA table**, where ANOVA stands for analysis of variance, see Table 2.2.

TABLE 2.2: ANOVA table.

Source	df	Sum of Squares	Mean Squares	F-ratio
Treatment	$g-1$	SS_{Trt}	$MS_{\text{Trt}} = \frac{SS_{\text{Trt}}}{g-1}$	$\frac{MS_{\text{Trt}}}{MS_E}$
Error	$N-g$	SS_E	$MS_E = \frac{SS_E}{N-g}$	

The mean squares (MS) are sum of squares (SS) that are normalized with the corresponding degrees of freedom. This is a so-called **one-way analysis of variance**, or **one-way ANOVA**, because there is only *one* factor involved. The corresponding cell means model is also called **one-way ANOVA model**.

Another rule of thumb for getting the degrees of freedom for the error is as follows: Total sum of squares has $N - 1$ degrees of freedom (we have N observations that fluctuate around the global mean), $g - 1$ degrees of freedom are "used" for the treatment effect (g groups minus one side constraint). Hence, the remaining part (the error) has $N - 1 - (g - 1) = N - g$ degrees of freedom, or as before, "total number of observations minus number of groups". A classical rule of thumb is that the error degrees of freedom should be at least 10; otherwise, the experiment has typically low power, because there are too many parameters in the model compared to the number of observations.

If all groups share the same expected value, the treatment sum of squares is typically small. Just due to the random nature of the response, small differences arise between the different (empirical) group means. The idea is now to compare this variation *between* groups with the variation *within* groups. If the variation between groups is substantially larger than the variation within groups, we have evidence against the null hypothesis.

In fact, it can be shown that under H_0 (single mean model) it holds that

$$F\text{-ratio} = \frac{MS_{\text{Trt}}}{MS_E} \sim F_{g-1,\, N-g}$$

where we denote by $F_{n,m}$ the so-called F-distribution with n and m degrees of freedom.

The F-distribution has two degrees of freedom parameters, one from the numerator, here, $g - 1$, and one from the denominator, here, $N - g$. See Figure 2.1 for the density of the F-distribution for some combinations of n and m. Increasing the denominator degrees of freedom (m) shifts the mass in the tail (the probability for very large values) toward zero. The $F_{1,m}$-distribution is a special case that we already know; it is the square of a t_m-distribution, a t-distribution with m degrees of freedom.

As with any other statistical test, we reject the null hypothesis if the observed value of the F-ratio, our test statistics, lies in an "extreme" region of the corresponding F-distribution. Any deviation from H_0 will lead to a larger value of the treatment sum of squares

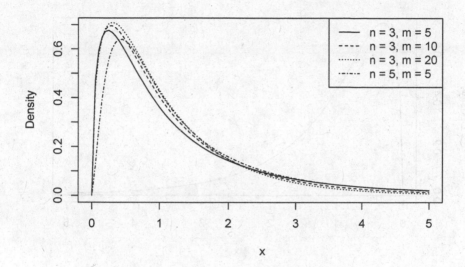

FIGURE 2.1: Densities of F-distributions.

(between groups) and to a larger F-ratio. Hence, unusually large values of the F-ratio provide evidence against H_0. Therefore, this is a one-sided test, which means that we reject H_0 in favor of H_A if the F-ratio is larger than the 95%-quantile of the corresponding $F_{g-1, N-g}$-distribution (if we use a significance level of 5%). We also use the notation $F_{n,m,\alpha}$ for the $(\alpha \times 100)$%-quantile of the $F_{n,m}$-distribution. Hence, we reject H_0 in favor of H_A if

$$F\text{-ratio} \geq F_{g-1, N-g, 1-\alpha}$$

if we use the significance level α, e.g., $\alpha = 0.05$. Alternatively, we can directly have a look at the p-value, typically automatically reported by software, and reject H_0 if the p-value is smaller than the chosen significance level. See Figure 2.2 for an illustration of the p-value for a situation where we have a realized F-ratio of 2.5. The p-value is simply the area under the curve for all values larger than 2.5 of the corresponding F-distribution.

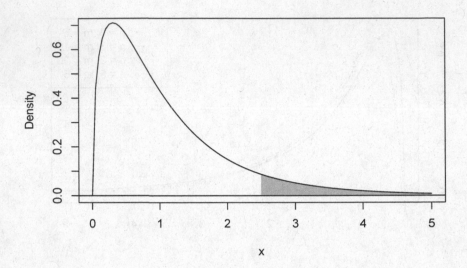

FIGURE 2.2: Illustration of the p-value (shaded area) for a realized F-ratio of 2.5.

As the test is based on the F-distribution, we simply call it an **F-test**. It is also called an **omnibus test** (Latin for "for all") or a **global test**, as it compares *all* group means simultaneously.

Note that the result of the F-test is independent of the side constraint that we use for the α_i's, as it only depends on the group means.

In R, we can use the summary function to get the ANOVA table and the p-value of the F-test.

```
summary(fit.plant)
```

```
##            Df Sum Sq Mean Sq F value Pr(>F)
## group       2   3.77   1.883    4.85  0.016
## Residuals  27  10.49   0.389
```

In the column labelled `Pr(>F)` we see the corresponding p-value: 0.016. Hence, we reject the null hypothesis on a 5% significance level.

Alternatively, we can also use the command `drop1` to perform the *F*-test. From a technical point of view, `drop1` tests whether all coefficients related to a "term" (here, a factor) are zero, which is equivalent to what we do with an *F*-test.

```
drop1(fit.plant, test = "F")
```

```
## Single term deletions
##
## Model:
## weight ~ group
##         Df Sum of Sq  RSS    AIC F value Pr(>F)
## <none>              10.5 -25.5
## group    2     3.77 14.3 -20.3    4.85  0.016
```

We get of course the same p-value as above, but `drop1` will be a useful function later (see Section 4.2.5).

As the *F*-test can also be interpreted as a test for comparing two different models, namely the cell means and the single means model, we have yet another approach in R. We can use the function `anova` to compare the two models.

```
## Fit single mean model (1 means global mean, i.e., intercept)
fit.plant.single <- aov(weight ~ 1, data = PlantGrowth)
## Compare with cell means model
anova(fit.plant.single, fit.plant)
```

```
## Analysis of Variance Table
##
## Model 1: weight ~ 1
## Model 2: weight ~ group
##   Res.Df  RSS Df Sum of Sq    F Pr(>F)
## 1     29 14.3
```

```
## 2      27 10.5  2       3.77 4.85  0.016
```

Again, we get the same p-value. We would also get the same p-value if we would use another side constraint, e.g., if we would use the `fit.plant2` object. If we use `anova` with only *one* argument, e.g., `anova(fit.plant)`, we get the same output as with `summary`.

To perform statistical inference for the *individual* α_i's we can use the commands `summary.lm` for tests and `confint` for confidence intervals.

```
summary.lm(fit.plant)
```

```
## ...
## Coefficients:
##              Estimate Std. Error t value Pr(>|t|)
## (Intercept)    5.032      0.197   25.53   <2e-16
## grouptrt1     -0.371      0.279   -1.33    0.194
## grouptrt2      0.494      0.279    1.77    0.088
## ...
```

```
confint(fit.plant)
```

```
##               2.5 % 97.5 %
## (Intercept)  4.62753  5.436
## grouptrt1   -0.94301  0.201
## grouptrt2   -0.07801  1.066
```

Now, interpretation of the output *highly* depends on the side constraint that is being used. For example, the confidence interval $[-0.943, 0.201]$ from the above output is a confidence interval for the difference between `trt1` and `ctrl` (because we used `contr.treatment`). Again, if we do not know the side constraint, we do not know what the estimates of the α_i's actually mean. If unsure, you can get the current setting with the command `options("contrasts")`.

2.2 Checking Model Assumptions

Statistical inference (p-values, confidence intervals, etc.) is only valid if the model assumptions are fulfilled. So far, this means:

- The errors are independent
- The errors are normally distributed
- The error variance is constant
- The errors have mean zero

The independence assumption is most crucial, but also most difficult to check. The randomization of experimental units to the different treatments is an important prerequisite (Montgomery, 2019; Lawson, 2014). If the independence assumption is violated, statistical inference can be very inaccurate. If the design contains some serial or spatial structure, some checks can be done as outlined below for the serial case.

In the ANOVA setting, the last assumption is typically not as important as in a regression setting. The reason is that we are typically fitting models that are "complex enough," and therefore do not show a lack of fit; see also the discussion in the following Section 2.2.1.

2.2.1 Residual Analysis

We *cannot* observe the errors ϵ_{ij} but we have the residuals r_{ij} as "estimates." Let us recall:

$$r_{ij} = y_{ij} - \hat{\mu}_i.$$

We now introduce different plots to check these assumptions. This means that we use graphical tools to perform *qualitative* checks.

Remark: Conceptually, one could also use statistical tests to perform these tasks. We prefer not to use them. Typically, these tests are again sensitive to model assumptions, e.g., tests for constant variance rely on the normal assumption. Or according to Box (1953),

> To make the preliminary test on variances is rather like putting to
> sea in a rowing boat to find out whether conditions are sufficiently
> calm for an ocean liner to leave port!

In addition, as with any other statistical test, power increases with
sample size. Hence, for large samples we get small p-values even
for very small deviations from the model assumptions. Even worse,
such a p-value does not tell us where the actual problem is, how
severe it is for the statistical inference and what could be done to
fix it.

We begin with a plot to check the normality assumption of the
error term.

QQ-Plot

In a **QQ-plot** we plot the empirical quantiles of the residuals or
"what we see in the data" vs. the theoretical quantiles or "what we
expect from the model". By default, a standard normal distribution
is the theoretical "reference distribution" (in such a situation, we
also call it a normal plot). The plot should show a more or less
straight line if the normality assumption is correct. In R, we can get
it with the function qqnorm or by calling plot on the fitted model
and setting which = 2, because in R, it is the second diagnostic plot.

```
plot(fit.plant, which = 2)
```

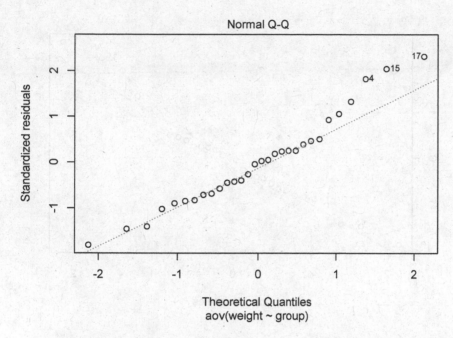

The plot for our data looks OK. As it might be difficult to judge what deviation of a straight line can still be tolerated, there are alternative implementations which also plot a corresponding "envelope" which should give you an idea of the natural variation. One approach is implemented in the function `qqPlot` of the package `car` (Fox and Weisberg, 2019):

```
library(car)
qqPlot(fit.plant, distribution = "norm")
```

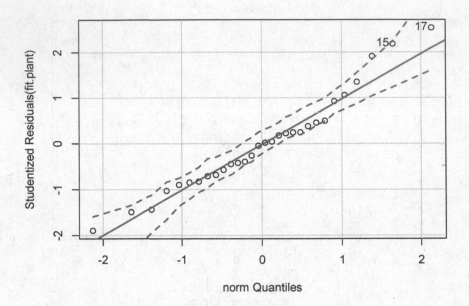

If the QQ-plot suggests non-normality, we can try to use a transformation of the response to accommodate this problem. Some
situations can be found in Figure 2.3. For a response which only
takes positive values, we can for example use the logarithm, the
square root (or any power less than one) if the residuals are skewed
to the right (Figure 2.3 middle). If the residuals are skewed to the
left (Figure 2.3 left), we could try a power greater than one. More
difficult is the situation where the residuals have a symmetric distribution but with heavier tails than the normal distribution (Figure
2.3 right). An approach could be to use methods that are presented
in Section 2.3. There is also the option to consider a whole family
of transformations, the so-called Box-Cox transformation, and to
choose the best fitting one, see Box and Cox (1964).

A transformation can be performed directly in the function call of
aov. While a generic call aov(y ~ treatment, data = data) would fit
a one-way ANOVA model with response y and predictor treatment
on the original scale of the response, we get with aov(log(y) ~
treatment, data = data) the one-way ANOVA model where we take

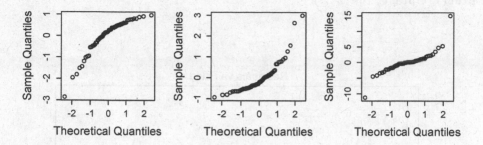

FIGURE 2.3: Examples of QQ-plots with deviations from normality. Left: Distribution is skewed to the left. Middle: Distribution is skewed to the right. Right: Distribution has heavier tails than the normal distribution.

the natural logarithm of the response. Of course you can also add a transformed variable yourself to the data-frame and use this variable in the model formula.

Note that whenever we transform the response, this comes at the price of a new interpretation of the parameters! Care has to be taken if one wants to interpret statistical inference (e.g., confidence intervals) on the original scale as many aspects unfortunately *do not* easily "transform back"! See Section 2.2.2 for more details.

Tukey-Anscombe Plot

The **Tukey-Anscombe plot (TA-plot)** plots the residuals r_{ij} vs. the fitted values $\hat{\mu}_i$ (estimated cell means). It allows us to check whether the residuals have constant variance. In addition, we could check whether there are regions where the model does not fit the data well such that the residuals do not have mean zero. However, this is not relevant yet as outlined below. In R, we can again use the function plot. Now we use the argument which = 1.

```
plot(fit.plant, which = 1)
```

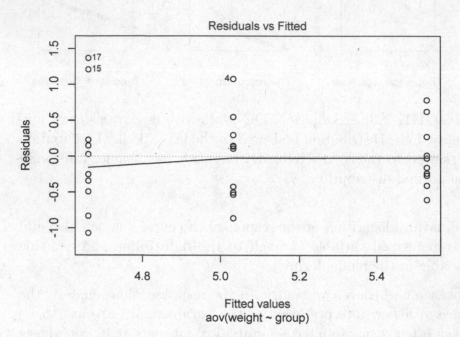

The solid line is a smoother (a local average) of the residuals. We could get rid of it by using the function call plot(fit.plant, which = 1, add.smooth = FALSE). For the one-way ANOVA situation, we could of course also read off the same information from the plot of the data itself. In addition, as we model one expected value per group, the model is "saturated," meaning it is the most complex model we can fit to the data, and the residuals automatically always have mean zero per group. The TA-plot will therefore be more useful later if we have more than one factor, see Chapter 4.

If you know this plot already from the regression setup and are worried about the vertical stripes, they are completely natural. The reason is that we only have $g = 3$ estimated cell means ($\hat{\mu}_i$'s) here on the x-axis.

In practice, a *very* common situation is that variation of a (positive) response is rather multiplicative (like $\pm 5\%$) than additive (like ±constant value), especially for situations where the response spreads over several orders of magnitude. The TA-plot then shows a linear increase of the standard deviation of the residuals (spread in the y-direction) with respect to the mean (on the x-axis), i.e., we have a "funnel shape" as in Figure 2.4 (left). Again, we can transform the response to fix this problem. For a situation like this, the logarithm will be the right choice, as can be seen in Figure 2.4 (right).

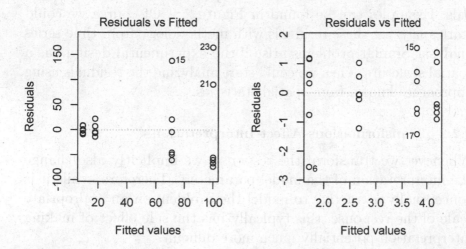

FIGURE 2.4: Tukey-Anscombe plot for the model using the original response (left) and for the model using the log-transformed response (right). Original data not shown.

The general rule is that if the standard deviation of the residuals is a monotone function of the fitted values (cell means), this can typically be fixed by a transformation of the response (as just seen with the logarithm). If the variance does not follow such a pattern, weights can be introduced to incorporate this so-called heteroscedasticity. The corresponding argument in the function aov is weights. However, we will not discuss this further here.

Index Plot

If the data has some serial structure, i.e., if observations were recorded in a certain time order, we typically want to check whether residuals close in time are more similar than residuals far apart, as this would be a violation of the independence assumption. We can do so by using a so-called **index plot** where we plot the residuals against time. For positively dependent residuals, we would see time periods where most residuals have the same sign, while for negatively dependent residuals, the residuals would "jump" too often from positive to negative compared to independent residuals. Examples can be found in Figure 2.5. Of course we could start analyzing these residuals with methodology from time-series analysis. Similar problems arise if the experimental design has a spatial structure. Then we could start analyzing the residuals using approaches known from spatial statistics.

2.2.2 Transformations Affect Interpretation

Whenever we transform the response, we implicitly also change the interpretation of the model parameters. Therefore, while it is conceptually attractive to model the problem on an appropriate scale of the response, this typically has the side effect of making interpretation potentially much more difficult.

Consider an example where we have a positive response. We transform it using the logarithm and use the standard model on the log-scale, i.e.,

$$\log(Y_{ij}) = \mu + \alpha_i + \epsilon_{ij}. \tag{2.6}$$

All the α_i's and their estimates have to be interpreted on the log-scale. Say we use the reference group encoding scheme (contr.treatment) where the first group is the reference group and we have $\hat{\alpha}_2 = 1.5$. This means that on the log-scale we estimate that the average value of group 2 is 1.5 larger than the average value of group 1 (additive shift). What about the effect on the *original* scale? On the log-scale we had $E[\log(Y_{ij})] = \mu + \alpha_i$. What does this tell us about $E[Y_{ij}]$, the expected value on the *original* scale? In general, the expected value does *not* directly follow a

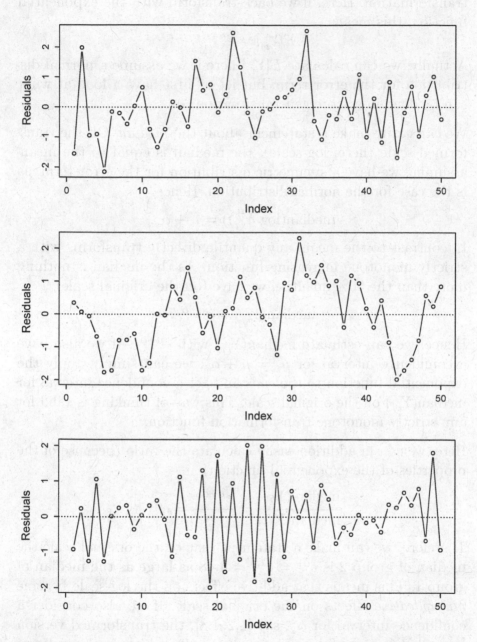

FIGURE 2.5: Examples of index plots: independent (top), positively correlated (middle) and negatively correlated residuals (bottom).

transformation. Here, if we back-transform with the exponential function, this means

$$E[Y_{ij}] \neq e^{\mu + \alpha_i}.$$

Actually, we can calculate $E[Y_{ij}]$ here if we assume a normal distribution for the error term, but let us first have a look at what can be done in a more general situation.

We can easily make a statement about the *median*. On the transformed scale (here, log-scale), the median is equal to the mean, assuming we have a symmetric distribution for the error term, as is the case for the normal distribution. Hence,

$$\text{median}(\log(Y_{ij})) = \mu + \alpha_i.$$

In contrast to the mean, any quantile directly transforms with a strictly monotone increasing function. As the median is nothing more than the 50% quantile, we have (on the original scale)

$$\text{median}(Y_{ij}) = e^{\mu + \alpha_i}.$$

Hence, we can estimate $\text{median}(Y_{ij})$ with $e^{\widehat{\mu} + \widehat{\alpha}_i}$. If we also have a confidence interval for $\mu_i = \mu + \alpha_i$, we can simply apply the exponential function to the ends and get a confidence interval for $\text{median}(Y_{ij})$ on the original scale. This way of thinking is valid for any strictly monotone transformation function.

Here, we can in addition easily calculate the ratio (because of the properties of the exponential function)

$$\frac{\text{median}(Y_{2j})}{\text{median}(Y_{1j})} = \frac{e^{\mu + \alpha_2}}{e^{\mu}} = e^{\alpha_2}.$$

Therefore, we can make a statement that on the original scale the median of group 2 is $e^{\widehat{\alpha}_2} = e^{1.5} = 4.48$ as large as the median of group 1. This means that additive effects on the log-scale become *multiplicative* effects on the original scale. If we also consider a confidence interval for α_2, say $[1.2, 1.8]$, the transformed version $[e^{1.2}, e^{1.8}]$ is a confidence interval for e^{α_2} which is the *ratio* of medians on the original scale, that is for (see above)

$$\frac{\text{median}(Y_{2j})}{\text{median}(Y_{1j})}.$$

Statements about the median on the original scale are typically very useful, as the distribution is skewed on the original scale. Hence, the median is a better, more interpretable, summary than the mean.

What can we do if we are still interested in the mean? If we use the standard assumption of normal errors with variance σ^2 in Equation (2.6), one can show that

$$E[Y_{ij}] = e^{\mu + \alpha_i + \sigma^2/2}.$$

Getting a confidence interval directly for $E[Y_{ij}]$ would need more work. However, if we consider the ratio of group 2 and group 1 (still using the reference group encoding scheme), we have

$$\frac{E[Y_{2j}]}{E[Y_{1j}]} = \frac{e^{\mu + \alpha_2 + \sigma^2/2}}{e^{\mu + \sigma^2/2}} = e^{\alpha_2}.$$

This is the very same result as above. Hence, the transformed confidence interval $[e^{1.2}, e^{1.8}]$ is also a confidence interval for the *ratio* of the mean values on the original scale! This "universal" quantification of the treatment effect on the original scale is a special feature of the log-transformation (because differences on the transformed scale are ratios on the original scale). When considering other transformation, this nice property is typically not available anymore. This makes interpretation much more difficult.

2.2.3 Checking the Experimental Design and Reports

When analyzing someone else's data or when reading a research paper, it is always a good idea to critically think about how the experiment was set up and how the statements in the research paper have to be understood:

- Is the data coming from an experiment or is it observational data? What claims are being made (causation or association)?
- In case of an experimental study: Was randomization used to assign the experimental units to the treatments? Might there be confounding variables which would weaken the results of the analysis?

- What is the experimental unit? Remember: When the treatment is allocated and administered to, e.g., feeders which are used by multiple animals, an experimental unit is given by a feeder and not by an individual animal. An animal would be an observational unit in this case. The average value across all animals of the same feeder would be what we would use as Y_{ij} in the one-way ANOVA model in Equation (2.4).
- When the estimated coefficients of treatment effects (the $\hat{\alpha}_i$'s) are being discussed: Is it clear what the corresponding side constraint is?
- If the response was transformed: Are the effects transformed appropriately?
- Was a residual analysis performed?

2.3 Nonparametric Approaches

What can we do if the distributional assumptions cannot be fulfilled? In the same way that there is a nonparametric alternative for the two-sample t-test (the so-called Mann-Whitney test, implemented in function `wilcox.test`), there is a non-parametric alternative for the one-way ANOVA setup: the **Kruskal-Wallis test**. The assumption of the Kruskal-Wallis test is that under H_0, all groups share the same distribution (*not* necessarily normal) and under H_A, at least one group has a shifted distribution (only location shift; the *shape* has to be the same). Instead of directly using the response, the Kruskal-Wallis test uses the corresponding ranks (we omit the details here). The test is available in the function `kruskal.test`, which can basically be called in the same way as `aov`:

```
fit.plant.kw <- kruskal.test(weight ~ group, data = PlantGrowth)
fit.plant.kw
```

```
##
##  Kruskal-Wallis rank sum test
##
```

```
## data:  weight by group
## Kruskal-Wallis chi-squared = 8, df = 2, p-value =
## 0.02
```

On this data set, the result is very similar to the parametric approach of aov. Note, however, that we only get the p-value of the global test and cannot do inference for individual treatment effects as was the case with aov.

A more general approach is using **randomization tests** where we would reshuffle the treatment assignment on the given data set to derive the distribution of some test statistics under the null hypothesis from the data itself. See for example Edgington and Onghena (2007) for an overview.

2.4 Power or "What Sample Size Do I Need?"

2.4.1 Introduction

By construction, a statistical test controls the so-called type I error rate with the significance level α. This means that the probability that we *falsely* reject the null hypothesis H_0 is less than or equal to α. Besides the type I error, there is also a type II error. It occurs if we *fail* to reject the null hypothesis even though the alternative hypothesis H_A holds. The probability of a type II error is typically denoted by β (and we are *not* controlling it). Note that there is no "universal" β, it actually depends on the specific alternative H_A that we believe in, i.e., assume. There is no such thing as "the" alternative H_A, we have to make a decision here.

The **power** of a statistical test, for a certain parameter setting under the alternative, is defined as

$P(\text{reject } H_0 \text{ given that a certain setting under } H_A \text{ holds}) = 1 - \beta.$

It translates as "what is the probability to get a significant result, assuming a certain parameter setting under the alternative H_A

holds". Intuitively, it seems clear that the "further away" we choose the parameter setting from H_0, the *larger* will be the power, or the *smaller* will be the probability of a type II error.

Calculating power is like a "thought experiment." We do *not* need data but a *precise* specification of the parameter setting under the alternative that we believe in ("what would happen if ...?"). This does not only include the parameters of interest but also nuisance parameters like the error variance σ^2.

2.4.2 Calculating Power for a Certain Design

Why should we be interested in such an abstract concept when planning an experiment? Power can be thought of as the probability of "success," i.e., getting a significant result. If we plan an experiment with low power, it means that we waste time and money because with high probability we are not getting a significant result. A rule of thumb is that power should be larger than 80%.

A typical question is: "What sample size do I need for my experiment?". There is of course no universal answer, but rather: "It depends on what power you like." So what does power actually depend on? It depends on

- design of the experiment (balanced, unbalanced, without blocking, with blocking, etc.)
- significance level α
- parameter setting under the alternative (incl. error variance σ^2)
- sample size n

We can mainly maximize power using the first and the last item from above.

Remark: In the same spirit, we could also ask ourselves how *precise* our parameter estimates will be. This is typically answered by checking if the widths of the corresponding confidence intervals (depending on the same quantities as above) are narrow enough.

For "easy" designs like a balanced completely randomized design, there are formulas to calculate power. They depend on the so-called non-central F-distribution which includes a non-centrality parameter (basically measuring how far away a parameter setting is from the null hypothesis of no treatment effect). We will not do these details here. As soon as the design is getting more complex as in the following chapters, things are typically much more complicated. Luckily, there is a simple way out. What we can always do is to *simulate* a lot of data sets under the alternative that we believe in and check how often we are rejecting the corresponding null hypothesis. The empirical rejection rate is then an estimate of the power; we can always increase the precision of this estimate by increasing the number of simulation runs.

We have a look at both approaches using a one-way ANOVA model with five groups (that is, we have $\mu_1, \mu_2, \ldots, \mu_5$) using a balanced design. The null hypothesis is

$$H_0 : \mu_1 = \ldots = \mu_5$$

while the alternative hypothesis is

$$H_A : \mu_k \neq \mu_l \text{ for at least one pair } k \neq l.$$

For example, we could assume that under H_A we have

$$\mu_1 = 57, \ \mu_2 = 63, \ \mu_3 = 60, \ \mu_4 = 60, \ \mu_5 = 60.$$

In addition, we have to specify the error variance. We assume it to be $\sigma^2 = 7$.

For such an easy design, we can use the function `power.anova.test` to calculate the power. It needs the number of groups as input (here, 5), the variance between the group means μ_i, the error variance σ^2 and the sample size within each group (assuming a balanced design). By default it uses a 5% significance level. If we assume $n = 4$ observations in each group, we have the following function call.

```
mu      <- c(57, 63, rep(60, 3))
sigma2 <- 7
power.anova.test(groups = length(mu), n = 4, between.var = var(mu),
                 within.var = sigma2)
```

```
##
##        Balanced one-way analysis of variance power calculation
##
##            groups = 5
##                 n = 4
##       between.var = 4.5
##        within.var = 7
##         sig.level = 0.05
##             power = 0.578
##
## NOTE: n is number in each group
```

According to the output, power is 58%. This means that we have a 58% chance to get a significant result under the above settings.

We can also leave away the argument n and use the argument power to get the required sample size (per group) for a certain power (here, 80%).

```
power.anova.test(groups = length(mu), between.var = var(mu),
                 within.var = sigma2, power = 0.8)
```

```
##
##        Balanced one-way analysis of variance power calculation
##
##            groups = 5
##                 n = 5.676
##       between.var = 4.5
##        within.var = 7
##         sig.level = 0.05
##             power = 0.8
##
## NOTE: n is number in each group
```

This means that for achieving a power of more than 80%, we need six observations per group.

What happens if we are facing a situation where there is no dedicated function like power.anova.test? We can easily calculate power using the simulation-based approach described above. Based on our specific setting under the alternative, we simulate many times a new data set, fit the one-way ANOVA model and check whether the corresponding F-test is significant.

```r
n.sim   <- 1000                    ## number of simulations
mu      <- c(57, 63, rep(60, 3))   ## group means
sigma2  <- 7                       ## error variance
n       <- 4                       ## number of observations per group

g       <- length(mu)
group   <- factor(rep(LETTERS[1:g], each = n))

results <- numeric(n.sim) ## vector to store results in

for(i in 1:n.sim){
  ## Simulate new response, build data set
  y <- rnorm(n * g, mean = rep(mu, each = n), sd = sqrt(sigma2))
  data <- data.frame(y = y, group = group)

  ## Fit one-way ANOVA model
  fit   <- aov(y ~ group, data = data)

  ## Extract result of global F-test
  results[i] <- summary(fit)[[1]][1, "Pr(>F)"] < 0.05 ## 1 = reject
}

mean(results) ## proportion of simulation runs that rejected H_0
```

```
## [1] 0.599
```

We get the result that in about 60% of the simulation runs, we got a significant result. This is very close to the result from above. If we wanted to increase the precision of our calculations, we could easily do this by using a larger value for the number of simulation runs n.sim.

A nice side effect of doing a power analysis is that you actually do the whole data analysis on simulated data and you immediately see whether this works as intended.

From a conceptual point of view, we can use such a simulation-based procedure for any design. Some implementations can be found in package Superpower (Lakens and Caldwell, 2021) and simr (Green and MacLeod, 2016). However, the number of parameters grows rather quickly with increasing model complexity (see the following chapters) and some of them are hard to specify in practice, e.g., the error variance, or the variances of different random effects as we will learn about in Chapter 6. If we are lucky, we have data from a pre-study. Unfortunately, the variance estimates are quite imprecise if we only have very limited amount of data. We could then consider a conservative estimate.

In that sense, the results of a power analysis are typically not very precise. However, they should still give us a *rough* idea about the required sample size in the sense of whether we need 6 or 60 observations per group.

In practice, resources (time, money, etc.) are limited. A power analysis still gives us an answer about whether it is actually worth doing the experimental study with the sample size that we can afford or whether it is a waste of resources (if the corresponding power is too low).

2.5   Adjusting for Covariates

Quite often, we do not only have the information about the treatment for each experimental unit, but also additional covariates (predictors), see also Section 1.2.1. Think for example of age, weight or some blood measurement when comparing the efficacy of drugs with patients. If these variables also have an influence on the response, we can get more precise, and less biased, estimates of the treatment effect when considering them. We have already seen this idea in Section 1.2.2 when discussing the basic idea of blocking.

However, if we for example do not have the information about these covariates early enough, we cannot create blocks. Nevertheless, we can still incorporate the corresponding information in the analysis. This is also called an **analysis of covariance (ANCOVA)**. From a technical point of view it is nothing more than a regression model (see also Section 2.6.2) with a categorical predictor, given by the treatment factor, and one (or multiple) typically continuous predictors.

Note that the additional covariates are *not* allowed to be affected by the treatment; otherwise, we have to be very careful of what the treatment effect actually means from a causal point of view. An example where this assumption is (trivially) fulfilled is the situation where the covariates are measured *before* the treatment is being applied.

Let us consider an example where we have a completely randomized design with a response y, a treatment factor treatment (with levels drug.A, drug.B and drug.C) and a continuous predictor x (e.g., think of some blood measurement). Our main interest is in the global test of the treatment factor.

We first load the data, inspect it and do a scatter plot which reveals that the covariate x is indeed predictive for the response y.

```
book.url <- "http://stat.ethz.ch/~meier/teaching/book-anova"
ancova <- read.table(file.path(book.url, "data/ancova.dat"),
                     header = TRUE, stringsAsFactors = TRUE)
str(ancova)
```

```
## 'data.frame':    60 obs. of  3 variables:
##  $ treatment: Factor w/ 3 levels "drug.A","drug.B",..: 1 1..
##  $ x        : num  21 17.8 24.6 ...
##  $ y        : num  48.9 39.1 67.3 ...
```

```
library(ggplot2)
ggplot(ancova, aes(x, y, shape = treatment)) + geom_point() +
  theme_bw()
```

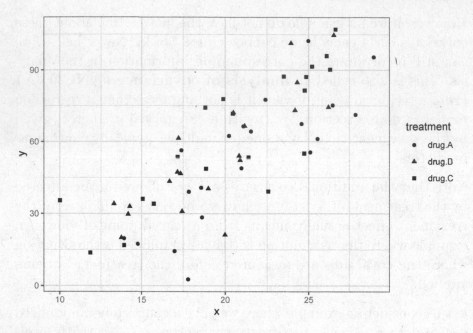

If we assume that the effect of the covariate is *linear* and the *same* for all treatment groups, we can extend the model in Equation (2.4) to include the covariate and get

$$Y_{ij} = \mu + \alpha_i + \beta \cdot x_{ij} + \epsilon_{ij} \qquad (2.7)$$

with ϵ_{ij} i.i.d. $\sim N(0, \sigma^2)$ and $i = 1, 2, 3$, $j = 1, \ldots, 20$. Note that x_{ij} is the value of the covariate of the jth experimental unit in treatment group i and β is the corresponding slope parameter. This means the model in Equation (2.7) fits three lines with different intercepts, given by $\mu + \alpha_i$, and a *common* slope parameter β, i.e., the lines are all parallel. This also means that the treatment effects (α_i's) are always the same. No matter what the value of the covariate, the difference between the treatment groups is always the same (given by the different intercepts). The fitted model is visualized in Figure 2.6.

How can we fit this model in R? We can still use the aov function, but we have to adjust the model formula to y ~ treatment + x (we could also use the lm function). We use the drop1 function to get the p-value for the global test of treatment which is adjusted for the covariate x (this will be discussed in more detail in Section 4.2.5).

```
options(contrasts = c("contr.treatment", "contr.poly"))
fit.ancova <- aov(y ~ treatment + x, data = ancova)
drop1(fit.ancova, test = "F")
```

```
## Single term deletions
##
## Model:
## y ~ treatment + x
##             Df Sum of Sq   RSS AIC F value  Pr(>F)
## <none>                   12337 328
## treatment  2      1953 14290 332    4.43   0.016
## x          1     27431 39768 396  124.51 7.4e-16
```

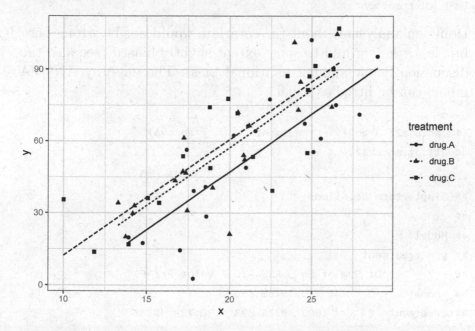

FIGURE 2.6: Scatter plot and illustration of the fitted model which includes the covariate x.

Note that we only lose one degree of freedom if we incorporate the covariate x into the model because it only uses one parameter (slope β). The treatment effect is significant. We cannot directly read off the estimate $\hat{\beta}$ from the summary output, but we can get it when calling `coef` (or `dummy.coef`).

```
coef(fit.ancova)
```

```
##     (Intercept) treatmentdrug.B treatmentdrug.C
##          -49.68           10.13           13.52
##               x
##            4.84
```

The estimate of the slope parameter can be found under x, hence $\hat{\beta} = 4.84$. However, note that the focus is clearly on the test of treatment. The covariate x is just a "tool" to get a more powerful test for treatment.

Doing an analysis without the covariate would not be wrong here, but less *efficient* and to some extent slightly biased (see also the discussion below about conditional bias). The one-way ANOVA model can be fitted as usual.

```
fit.ancova2 <- aov(y ~ treatment, data = ancova)
drop1(fit.ancova2, test = "F")
```

```
## Single term deletions
##
## Model:
## y ~ treatment
##           Df Sum of Sq   RSS AIC F value Pr(>F)
## <none>                 39768 396
## treatment  2     1002 40770 393    0.72   0.49
```

We observe that the p-value of treatment is much larger compared to the previous analysis. The reason is that there is much more unexplained variation. In the classical one-way ANOVA model (stored in object fit.ancova2), we have to deal with the whole variation of the response. In Figure 2.6 this is the complete variation in the direction of the y-axis (think of projecting all points on the *y*-axis). Alternatively, we can also visualize this by individual boxplots.

```
boxplot(y ~ treatment, data = ancova)
```

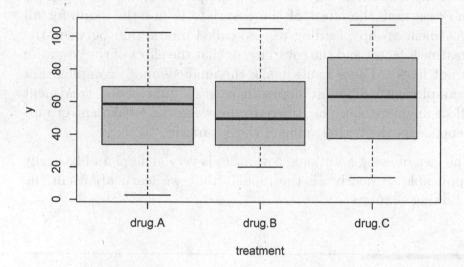

With the ANCOVA approach, we have the predictor x which explains, and therefore removes, a lot of the variation and what is left for the error term is only the variation around the straight lines in Figure 2.6.

Here are a few words about interpretation. The treatment effects that we estimate with the ANCOVA model in Equation (2.7) are *condititional* treatment effects (conditional on the same value of the covariate x). If we do a completely randomized design and ignore the covariate by using "only" a one-way ANOVA model, we get unbiased estimates of the treatment effect, in the sense that if we repeat this procedure many times and take the sample mean of the estimates, we get the true values. However, for a given realization, the usual ANOVA estimate can be slightly biased because the covariate is not perfectly balanced between the treatment groups. This is also called a *conditional* bias. This bias is typically small for a completely randomized design. Hence, in these situations, using the covariate is mainly due to efficiency gains, i.e., power. However, for observational data, the covariate imbalance can be substantial, and using a model including the covariate is typically (desperately) needed to reduce bias. See also the illustrations in section 6.4 of Huitema (2011).

This was of course only an easy example. It could very well be the case that the effect of the covariate is not the same for all treatment groups, leading to a so-called interaction between the treatment factor and the covariate, or that the effect of the covariate is not linear. These issues make the analysis more complex. For example, with different slopes there is no "universal" treatment effect anymore, but the difference between the treatment groups depends on the actual value of the covariate.

The idea of using additional covariates is very general and basically applicable to nearly all the models that we learn about in the following chapters.

2.6   Appendix

2.6.1   Ordered Factors: Polynomial Encoding Scheme

To illustrate the special aspects of an ordered factor, we consider the following hypothetical example. The durability of a technical component (variable `durability` taking values between 0 and 100) was tested for different pressure levels applied in the production process (variable `pressure` measured in bar). Note that the predictor `pressure` could also be interpreted as a continuous variable here. If we treat it as a factor, we are in the situation to have an *ordered* factor.

We first load the data and make sure that pressure is encoded as an *ordered* factor.

```
book.url <- "http://stat.ethz.ch/~meier/teaching/book-anova"
stress <- read.table(file.path(book.url, "data/stress.dat"),
                     header = TRUE)
stress[,"pressure"] <- ordered(stress[,"pressure"])
str(stress)
```

```
## 'data.frame':    16 obs. of  2 variables:
## $ pressure  : Ord.factor w/ 4 levels "1"<"1.5"<"2"<..: 1 ..
## $ durability: int  54 58 65 58 77 67 ...
```

A visualization of the data can be found in Figure 2.7 (R code not shown).

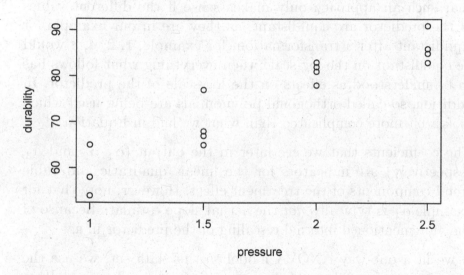

FIGURE 2.7: Visualization of durability vs. pressure.

Due to the ordered nature of `pressure`, it makes sense to also consider other parametrizations than those presented in Table 2.1. The 4 different cell means uniquely define a polynomial of order 3 (because 2 points define a linear function, 3 points define a quadratic function, 4 points define a polynomial of order 3). Hence, for the general situation with g treatments, we can use a polynomial of order $g - 1$ to parametrize the g different cell means. Here, this means

$$\mu_i = \mu + \alpha_1 \cdot \text{pressure}_i + \alpha_2 \cdot \text{pressure}_i^2 + \alpha_3 \cdot \text{pressure}_i^3. \quad (2.8)$$

Think of simply plugging in the actual numerical value for pressure$_i$ in Equation (2.8).

How do we tell R to use such an approach? The second argument when calling `options(contrasts = c("contr.sum", "contr.poly"))` is the parametrization for ordered factors. The default value is `"contr.poly"` and basically does what we described above. Unfortunately, the reality is a bit more complicated because R internally uses the value 1 for the smallest level, 2 for the second smallest up to (here) 4 for the largest level (originally 2.5 bar). This means that such an approach only makes sense if the different values of the predictor are equidistant, as they are in our example, or equidistant after a transformation, for example, 1, 2, 4, 8 would be equidistant on the log-scale; then everything what follows has to be understood as effects on the log-scale of the predictor. In addition, so-called orthogonal polynomials are being used which look a bit more complicated than what we had in Equation (2.8).

The coefficients that we see later in the output ($\hat{\alpha}_1, \hat{\alpha}_2$ and $\hat{\alpha}_3$, respectively) are indicators for the linear, quadratic and cubic trend components of the treatment effect. However, note that for example $\hat{\alpha}_1$ is typically *not* the actual slope estimate because of the aforementioned internal rescaling of the predictor in R.

If we fit a one-way ANOVA model to this data set, we use the same approach as always (remember: we now simply use another fancy parametrization for the cell means which does not affect the F-test).

```
fit.stress <- aov(durability ~ pressure, data = stress)
summary(fit.stress)
```

```
##          Df Sum Sq Mean Sq F value  Pr(>F)
```

```
## pressure      3    1816     605      37 2.4e-06
## Residuals    12     196      16
```

The effect of `pressure` is highly significant. Now let us have a look at the statistical inference for the individual α_i's.

```
summary.lm(fit.stress)
```

```
## ...
## Coefficients:
##               Estimate Std. Error t value Pr(>|t|)
## (Intercept)      73.81       1.01   73.01  < 2e-16
## pressure.L       21.07       2.02   10.42  2.3e-07
## pressure.Q       -2.63       2.02   -1.30     0.22
## pressure.C       -1.73       2.02   -0.86     0.41
## ...
```

Now the coefficient names are different than what we used to get. The three rows labelled with `pressure.L`, `pressure.Q` and `pressure.C` are the estimates of the linear, quadratic and cubic component, i.e., $\hat{\alpha}_1$, $\hat{\alpha}_2$, $\hat{\alpha}_3$. We shouldn't spend too much attention on the actual estimates (column `Estimate`) because the value depends on the internal parametrization. Luckily, the p-values still offer an easy interpretation. For example, the null hypothesis of the row `pressure.L` is: "the linear component of the relationship between `durability` and `pressure` is zero," or equivalently $H_0 : \alpha_1 = 0$.

From the output, we can conclude that only the linear part is significant with a very small p-value. Hence, the conclusion would be that the data can be described using a linear trend (which could also be used to interpolate between the observed levels of `pressure`), and there is no evidence of a quadratic or cubic component. In that sense, such an ANOVA model is very close to an approach using a linear regression model where we would treat `pressure` as a continuous predictor. See also Section 2.6.2 for more information that ANOVA models are basically nothing more than special linear regression models.

2.6.2 Connection to Regression

We can also write Equation (2.4) in matrix notation

$$Y = X\beta + E,$$

where $Y \in \mathbb{R}^N$ contains all the response values, X is an $N \times g$ matrix, the so-called **design matrix**, $\beta \in \mathbb{R}^g$ contains all parameters and $E \in \mathbb{R}^N$ are the error terms. A *row* of the design matrix corresponds to an individual *observation*, while a *column* corresponds to a *parameter*. This is a typical setup for any regression problem.

We use a subset of the PlantGrowth data set (only the first two observations per group) for illustration.

```
PlantGrowth[c(1, 2, 11, 12, 21, 22),]
```

```
##     weight group
## 1     4.17  ctrl
## 2     5.58  ctrl
## 11    4.81  trt1
## 12    4.17  trt1
## 21    6.31  trt2
## 22    5.12  trt2
```

Hence, for the response we have the vector of observed values

$$y = \begin{pmatrix} 4.17 \\ 5.58 \\ 4.81 \\ 4.17 \\ 6.31 \\ 5.12 \end{pmatrix}.$$

How do the design matrix X and the parameter vector β look like? This depends on the side constraint on the α_i's. If we use `contr.treatment`, we get

$$X = \begin{pmatrix} 1 & 0 & 0 \\ 1 & 0 & 0 \\ 1 & 1 & 0 \\ 1 & 1 & 0 \\ 1 & 0 & 1 \\ 1 & 0 & 1 \end{pmatrix}, \qquad \beta = \begin{pmatrix} \mu \\ \alpha_2 \\ \alpha_3 \end{pmatrix}$$

because the group `ctrl` is the reference level (hence $\alpha_1 = 0$). For `contr.sum` we have

$$X = \begin{pmatrix} 1 & 1 & 0 \\ 1 & 1 & 0 \\ 1 & 0 & 1 \\ 1 & 0 & 1 \\ 1 & -1 & -1 \\ 1 & -1 & -1 \end{pmatrix}, \qquad \beta = \begin{pmatrix} \mu \\ \alpha_1 \\ \alpha_2 \end{pmatrix}$$

because $\alpha_3 = -(\alpha_1 + \alpha_2)$.

Hence, what we do is nothing more than fitting a linear regression model with a categorical predictor. The categorical predictor is being represented by a *set* of dummy variables.

3

Contrasts and Multiple Testing

3.1 Contrasts

3.1.1 Introduction

The F-test is rather *unspecific*. It basically gives us a "Yes/No" answer to the question: "Is there any treatment effect at all?". It does not tell us what specific treatment or treatment combination is special. Quite often, we have a more specific question than the aforementioned *global* null hypothesis. For example, we might want to compare a set of new treatments vs. a control treatment or we want to do pairwise comparisons between many (or all) treatments.

Such kinds of questions can typically be formulated as a so-called **contrast**. Let us start with a toy example based on the PlantGrowth data set. If we only wanted to compare trt1 (μ_2) with ctrl (μ_1), we could set up the null hypothesis

$$H_0 : \mu_1 - \mu_2 = 0$$

vs. the alternative

$$H_A : \mu_1 - \mu_2 \neq 0.$$

We can encode this with a vector $c \in \mathbb{R}^g$

$$H_0 : \sum_{i=1}^{g} c_i \mu_i = 0. \tag{3.1}$$

In this example, we have $g = 3$ and the vector c is given by $c = (1, -1, 0)$, with respect to ctrl, trt1 and trt2. Hence, a contrast is nothing more than an encoding of our own specific research question. A more sophisticated example would be $c = (1, -1/2, -1/2)$ which compares ctrl vs. the average value of trt1 and trt2 which we would write as $H_0 : \mu_1 - \frac{1}{2}(\mu_2 + \mu_3) = 0$.

Typically, we have the side constraint

$$\sum_{i=1}^{g} c_i = 0$$

which ensures that the contrast is about *differences* between treatments and not about the overall level of the response.

A contrast can also be thought of as one-dimensional "aspect" of the multi-dimensional treatment effect, if we have $g > 2$ different treatments.

We estimate the corresponding true, but unknown, value $\sum_{i=1}^{g} c_i \mu_i$ (a linear combination of model *parameters*!) with

$$\sum_{i=1}^{g} c_i \hat{\mu}_i = \sum_{i=1}^{g} c_i \bar{y}_{i \cdot} .$$

In addition, we could derive its accuracy (standard error), construct confidence intervals and do tests. We omit the theoretical details and continue with our example.

In R, we use the function glht (general linear hypotheses) of the package multcomp (Hothorn et al., 2008). It uses the fitted one-way ANOVA model, which we refit here for the sake of completeness.

```
fit.plant <- aov(weight ~ group, data = PlantGrowth)
```

We first have to specify the contrast for the factor group with the function mcp (multiple comparisons; for the moment we only consider a single test here) and use the corresponding output as argument linfct (linear function) in glht. All these steps together look as follows:

```
library(multcomp)
plant.glht <- glht(fit.plant,
                   linfct = mcp(group = c(1, -1/2, -1/2)))
summary(plant.glht)
```

```
## ...
## Linear Hypotheses:
##        Estimate Std. Error t value Pr(>|t|)
## 1 == 0  -0.0615     0.2414   -0.25      0.8
## ...
```

This means that we estimate the difference between ctrl and the average value of trt1 and trt2 as -0.0615 and we are not rejecting the null hypothesis because the p-value is large. The annotation 1 == 0 means that this line tests whether the first (here, and only) contrast is zero or not (if needed, we could also give a custom name to each contrast). We get a confidence interval by using the function confint.

```
confint(plant.glht)
```

```
## ...
## Linear Hypotheses:
##        Estimate lwr      upr
## 1 == 0 -0.0615  -0.5569  0.4339
```

Hence, the 95% confidence interval for $\mu_1 - \frac{1}{2}(\mu_2 + \mu_3)$ is given by $[-0.5569, 0.4339]$.

An alternative to package multcomp is package emmeans. One way of getting statistical inference for a contrast is by using the function contrast on the output of emmeans. The corresponding function call for the contrast from above is as follows:

```
library(emmeans)
plant.emm <- emmeans(fit.plant, specs = ~ group)
contrast(plant.emm, method = list(c(1, -1/2, -1/2)))
```

```
## contrast           estimate    SE df t.ratio p.value
## c(1, -0.5, -0.5)  -0.0615 0.241 27 -0.255  0.8009
```

A confidence interval can also be obtained by calling `confint` (not shown).

Remark: For ordered factors we could also define contrasts which capture the linear, quadratic or higher-order trend if applicable. This is in fact exactly what is being used when using `contr.poly` as seen in Section 2.6.1. We call such contrasts **polynomial contrasts**. The result can directly be read off the output of `summary.lm`. Alternatively, we could also use `emmeans` and set `method = "poly"` when calling the `contrast` function.

3.1.2   Some Technical Details

Every contrast has an associated **sum of squares**

$$SS_c = \frac{\left(\sum_{i=1}^{g} c_i \bar{y}_{i.}\right)^2}{\sum_{i=1}^{g} \frac{c_i^2}{n_i}}$$

having *one* degree of freedom. Hence, for the corresponding mean squares it holds that $MS_c = SS_c$. This looks unintuitive at first sight, but it is nothing more than the square of the t-statistic for the special model parameter $\sum_{i=1}^{g} c_i \mu_i$ with the null hypothesis defined in Equation (3.1) (without the MS_E factor). You can think of SS_c as the "part" of SS_{Trt} in "direction" of c.

Under $H_0 : \sum_{i=1}^{g} c_i \mu_i = 0$ it holds that

$$\frac{MS_c}{MS_E} \sim F_{1, N-g}.$$

Because $F_{1, m} = t_m^2$ (the square of a t_m-distribution with m degrees of freedom), this is nothing more than the "squared version" of the t-test.

Two contrasts c and c^* are called **orthogonal** if

$$\sum_{i=1}^{g} \frac{c_i c_i^*}{n_i} = 0.$$

If two contrasts c and c^* are orthogonal, the corresponding esti-mates are stochastically independent. This means that if we know something about one of the contrasts, this does *not* help us in making a statement about the other one.

If we have g treatments, we can find $g - 1$ different orthogonal con-trasts (one dimension is already used by the global mean $(1, \ldots, 1)$). A set of *orthogonal* contrasts **partitions** the treatment sum of squares meaning that if $c^{(1)}, \ldots, c^{(g-1)}$ are orthogonal contrasts it holds that

$$SS_{c^{(1)}} + \cdots + SS_{c^{(g-1)}} = SS_{\text{Trt}}.$$

Intuition: "We get all information about the treatment by asking the right $g - 1$ questions."

However, your research questions define the contrasts, not the orthogonality criterion!

3.2 Multiple Testing

The problem with all statistical tests is the fact that the overall type I error rate increases with increasing number of tests. Assume that we perform m independent tests whose null hypotheses we label with $H_{0,j}$, $j = 1, \ldots, m$. Each test uses an individual significance level of α. Let us first calculate the probability to make *at least one* false rejection for the situation where all $H_{0,j}$ are true. To do so, we first define the event $A_j = \{\text{test } j \text{ falsely rejects } H_{0,j}\}$.

The event "there is at least one false rejection among all m tests" can be written as $\cup_{j=1}^{m} A_j$. Using the complementary event and the independence assumption, we get

$$P\left(\bigcup_{j=1}^{m} A_j\right) = 1 - P\left(\bigcap_{j=1}^{m} A_j^c\right)$$

$$= 1 - \prod_{j=1}^{m} P(A_j^c)$$

$$= 1 - (1 - \alpha)^m.$$

Even for a small value of α, this is close to 1 if m is large. For example, using $\alpha = 0.05$ and $m = 50$, this probability is 0.92!

This means that if we perform many tests, we expect to find some significant results, even if *all* null hypotheses are true. Somehow we have to take into account the number of tests that we perform to control the overall type I error rate.

Using similar notation as Bretz et al. (2011), we list the potential outcomes of a total of m tests, among which m_0 null hypotheses are true, in Table 3.1.

TABLE 3.1: Outcomes of a total of m statistical tests, among which m_0 null hypotheses are true. Capital letters indicate random variables.

	H_0 true	H_0 false	Total
Significant	V	S	R
Not significant	U	T	W
Total	m_0	$m - m_0$	m

For example, V is the number of wrongly rejected null hypotheses (type I errors, also known as false positives), T is the number of type II errors (also known as false negatives), R is the number of significant results (or "discoveries"), etc.

Using this notation, the overall or **family-wise error rate (FWER)** is defined as the probability of rejecting *at least one* of the *true* H_0's:

$$\text{FWER} = P(V \geq 1).$$

The family-wise error rate is very strict in the sense that we are not considering the actual *number* of wrong rejections, we are just interested in whether there is *at least one*. This means the situation where we make (only) $V = 1$ error is equally "bad" as the situation where we make $V = 20$ errors.

We say that a procedure controls the family-wise error rate in the *strong sense* at level α if

$$\text{FWER} \leq \alpha$$

for *any* configuration of true and non-true null hypotheses. A typical choice would be $\alpha = 0.05$.

Another error rate is the **false discovery rate (FDR)** which is the expected fraction of false discoveries,

$$\text{FDR} = E\left[\frac{V}{R}\right].$$

Controlling FDR at, e.g., level 0.2 means that on average in our list of "significant findings" only 20% are not "true findings" (false positives). If we can live with a certain amount of false positives, the relevant quantity to control is the false discovery rate.

If a procedure controls FWER at level α, FDR is *automatically* controlled at level α too (Bretz et al., 2011). On the other hand, a procedure that controls FDR at level α might have a much larger error rate regarding FWER. Hence, FWER is a much stricter (more conservative) criterion leading to fewer rejections.

We can also control the error rates for confidence intervals. We call a set of confidence intervals **simultaneous confidence intervals** at level $(1 - \alpha)$ if the probability that *all* intervals cover the corresponding true parameter value is $(1 - \alpha)$. This means that we can look at all confidence intervals at the same time and get the correct "big picture" with probability $(1 - \alpha)$.

In the following, we focus on the FWER and simultaneous confidence intervals.

We typically start with individual p-values (the ordinary p-values corresponding to the $H_{0,j}$'s) and modify or adjust them such that the appropriate *overall* error rate (like FWER) is being controlled. Interpretation of an individual p-value is as you have learned in your introductory course ("the probability to observe an event as extreme as ..."). The modified p-values should be interpreted as the smallest *overall* error rate such that we can reject the corresponding null hypothesis.

The theoretical background for most of the following methods can be found in Bretz et al. (2011).

3.2.1 Bonferroni

The **Bonferroni** correction is a very generic but conservative approach. The idea is to use a more restrictive (individual) significance level of $\alpha^* = \alpha/m$. For example, if we have $\alpha = 0.05$ and $m = 10$, we would use an individual significance level of $\alpha^* = 0.005$. This procedure controls the FWER in the strong sense for *any* dependency structure of the different tests. Equivalently, we can also multiply the original p-values by m and keep using the original significance level α. Especially for large m, the Bonferroni correction is very conservative leading to low power.

Why does it work? Let M_0 be the index set corresponding to the true null hypotheses, with $|M_0| = m_0$. Using an individual significance level of α/m we get

$$P(V \geq 1) = P\left(\bigcup_{j \in M_0} \text{reject } H_{0,j}\right) \leq \sum_{j \in M_0} P(\text{reject } H_{0,j})$$
$$\leq m_0 \frac{\alpha}{m} \leq \alpha.$$

The confidence intervals based on the adjusted significance level are simultaneous (e.g., for $\alpha = 0.05$ and $m = 10$ we would need individual 99.5% confidence intervals).

We have a look at the previous example where we have two contrasts, $c_1 = (1, -1/2, -1/2)$ ("control vs. the average of the remaining treatments") and $c_2 = (1, -1, 0)$ ("control vs. treatment 1").

We first construct a contrast matrix where the two rows correspond to the two contrasts. Calling `summary` with `test = adjusted("none")` gives us the usual individual, i.e., unadjusted p-values.

```
library(multcomp)
## Create a matrix where each *row* is a contrast
K <- rbind(c(1, -1/2, -1/2), ## ctrl vs. average of trt1 and trt2
           c(1, -1, 0))        ## ctrl vs. trt1
plant.glht.K <- glht(fit.plant, linfct = mcp(group = K))

## Individual p-values
summary(plant.glht.K, test = adjusted("none"))
```

```
## ...
## Linear Hypotheses:
##        Estimate Std. Error t value Pr(>|t|)
## 1 == 0  -0.0615     0.2414   -0.25     0.80
## 2 == 0   0.3710     0.2788    1.33     0.19
## (Adjusted p values reported -- none method)
```

If we use `summary` with `test = adjusted("bonferroni")` we get the Bonferroni-corrected p-values. Here, this consists of a multiplication by 2 (you can also observe that if the resulting p-value is larger than 1, it will be set to 1).

```
## Bonferroni corrected p-values
summary(plant.glht.K, test = adjusted("bonferroni"))
```

```
## ...
## Linear Hypotheses:
##        Estimate Std. Error t value Pr(>|t|)
## 1 == 0  -0.0615     0.2414   -0.25     1.00
## 2 == 0   0.3710     0.2788    1.33     0.39
## (Adjusted p values reported -- bonferroni method)
```

By default, `confint` calculates simultaneous confidence intervals. Individual confidence intervals can be computed by setting the argument `calpha = univariate_calpha()`, critical value of α, in `confint` (not shown).

With emmeans, the function call to get the Bonferroni-corrected
p-values is as follows:

```
contrast(plant.emm, method = list(c(1, -1/2, -1/2), c(1, -1, 0)),
         adjust = "bonferroni")
```

```
##   contrast          estimate     SE df t.ratio p.value
##   c(1, -0.5, -0.5)   -0.0615 0.241 27 -0.255   1.0000
##   c(1, -1, 0)         0.3710 0.279 27  1.331   0.3888
##
## P value adjustment: bonferroni method for 2 tests
```

3.2.2   Bonferroni-Holm

The **Bonferroni-Holm** procedure (Holm, 1979) also controls the
FWER in the strong sense. It is less conservative and *uniformly*
more powerful, which means always better, than Bonferroni. It
works in the following *sequential* way:

1. Sort p-values from small to large: $p_{(1)} \leq p_{(2)} \leq \cdots \leq p_{(m)}$.
2. For $j = 1, 2, \ldots$: Reject null hypothesis if $p_{(j)} \leq \frac{\alpha}{m-j+1}$.
3. Stop when reaching the *first* non-significant p-value (and do *not*
   reject the remaining null hypotheses).

Note that only the *smallest* p-value has the traditional Bonferroni
correction. Bonferroni-Holm is a so-called *stepwise*, more precisely
step-down, procedure as it starts at the smallest p-value and steps
down the sequence of p-values (Bretz et al., 2011). Note that this
procedure only works with p-values but *cannot* be used to construct
confidence intervals.

With the multcomp package, we can set the argument test of the
function summary accordingly:

```
summary(plant.glht.K, test = adjusted("holm"))
```

```
## ...
## Linear Hypotheses:
```

```
##          Estimate Std. Error t value Pr(>|t|)
## 1 == 0   -0.0615     0.2414   -0.25    0.80
## 2 == 0    0.3710     0.2788    1.33    0.39
## (Adjusted p values reported -- holm method)
```

With `emmeans`, the argument `adjust = "holm"` has to be used (not shown). In addition, this is also implemented in the function `p.adjust` in R.

3.2.3 Scheffé

The **Scheffé** procedure (Scheffé, 1959) controls for the search over *any* possible contrast. This means we can try out as many contrasts as we like and still get honest p-values! This is even true for contrasts that are suggested by the data, which were not planned beforehand, but only after seeing some special structure in the data. The price for this nice property is low power.

The Scheffé procedure works as follows: We start with the sum of squares of the contrast SS_c. Remember: This is the part of the variation that is explained by the contrast, like a one-dimensional aspect of the multi-dimensional treatment effect. Now we are conservative and treat this as if it would be the *whole* treatment effect. This means we use $g - 1$ as the corresponding degrees of freedom and therefore calculate the mean squares as $SS_c/(g-1)$. Then we build the usual F-ratio by dividing through MS_E, i.e.,

$$\frac{SS_c/(g-1)}{MS_E}$$

and compare the realized value to an $F_{g-1,\,N-g}$-distribution (the same distribution that we would also use when testing the *whole* treatment effect).

Note: Because it holds that $SS_c \leq SS_{\text{Trt}}$, we do not even have to start searching if the F-test is not significant.

What we described above is equivalent to taking the "usual" F-ratio of the contrast (typically available from any software) and use the distribution $(g-1) \cdot F_{g-1,\,N-g}$ instead of $F_{1,\,N-g}$ to calculate the p-value.

We can do this manually in R with the multcomp package. We first treat the contrast as an "ordinary" contrast and then do a manual calculation of the p-value. As glht reports the value of the *t*-test, we first have to take the square of it to get the *F*-ratio. As an example, we consider the contrast $c = (1/2, -1, 1/2)$ (the mean of the two groups with large values vs. the group with small values, see Section 2.1.2).

```
plant.glht.scheffe <- glht(fit.plant,
                      linfct = mcp(group = c(1/2, -1, 1/2)))
## p-value according to Scheffe (g = 3, N - g = 27)
pf((summary(plant.glht.scheffe)$test$tstat)^2 / 2, 2, 27,
   lower.tail = FALSE)
```

```
##        1
## 0.05323
```

If we use a significance level of 5% we do not get a significant result, with the more extreme contrast $c = (0, -1, 1)$ we would be successful.

Confidence intervals can be calculated too by inverting the test from above, see Section 5.3 in Oehlert (2000) for more details.

```
summary.glht <- summary(plant.glht.scheffe)$test
estimate <- summary.glht$coefficients ## estimate
sigma    <- summary.glht$sigma          ## standard error
crit.val <- sqrt(2 * qf(0.95, 2, 27)) ## critical value
estimate + c(-1, 1) * sigma * crit.val
```

```
## [1] -0.007316  1.243316
```

An alternative implementation is also available in the function ScheffeTest of package DescTools (Signorell et al., 2021).

In emmeans, the argument adjust = "scheffe" can be used. For the same contrast as above, the code would be as follows (the argument scheffe.rank has to be set to the degrees of freedom of the factor, here 2).

```
summary(contrast(plant.emm, method = list(c(1/2, -1, 1/2)),
                 adjust = "scheffe"), scheffe.rank = 2)
```

```
## contrast          estimate    SE df t.ratio p.value
## c(0.5, -1, 0.5)      0.618 0.241 27 2.560    0.0532
##
## P value adjustment: scheffe method with rank 2
```

Confidence intervals can be obtained by replacing summary with confint in the previous function call (not shown).

3.2.4 Tukey Honest Significant Differences

A special case of a multiple testing problem is the comparison between *all* possible pairs of treatments. There are a total of $g \cdot (g - 1)/2$ pairs that we can inspect. We could perform all pairwise t-tests with the function pairwise.t.test; it uses a pooled standard deviation estimate from all groups.

```
## Without correction (but pooled sd estimate)
pairwise.t.test(PlantGrowth$weight, PlantGrowth$group,
                p.adjust.method = "none")
```

```
##
##  Pairwise comparisons using t tests with pooled SD
##
## data:  PlantGrowth$weight and PlantGrowth$group
##
##      ctrl  trt1
## trt1 0.194 -
## trt2 0.088 0.004
##
## P value adjustment method: none
```

The output is a *matrix* of p-values of the corresponding comparisons (see row and column labels). We could now use the Bonferroni correction method, i.e., `p.adjust.method = "bonferroni"` to get p-values that are adjusted for multiple testing.

```
## With correction (and pooled sd estimate)
pairwise.t.test(PlantGrowth$weight, PlantGrowth$group,
                p.adjust.method = "bonferroni")
```

```
##
##  Pairwise comparisons using t tests with pooled SD
##
## data:  PlantGrowth$weight and PlantGrowth$group
##
##      ctrl trt1
## trt1 0.58 -
## trt2 0.26 0.01
##
## P value adjustment method: bonferroni
```

However, there exists a better, more powerful alternative which is called **Tukey Honest Significant Differences (HSD)**. The balanced case goes back to Tukey (1949a), an extension to unbalanced situations can be found in Kramer (1956), which is also discussed in Hayter (1984). Think of a procedure that is custom tailored for the situation where we want to do a comparison between all possible pairs of treatments. We get both p-values (which are adjusted such that the family-wise error rate is being controlled) and simultaneous confidence intervals. In R, this is directly implemented in the function `TukeyHSD` and of course both packages `multcomp` and `emmeans` contain an implementation too.

We can directly call `TukeyHSD` with the fitted model as the argument:

```
TukeyHSD(fit.plant)
```

```
##   Tukey multiple comparisons of means
##     95% family-wise confidence level
##
## Fit: aov(formula = weight ~ group, data = PlantGrowth)
```

```
##
## $group
##               diff     lwr    upr  p adj
## trt1-ctrl  -0.371 -1.0622 0.3202 0.3909
## trt2-ctrl   0.494 -0.1972 1.1852 0.1980
## trt2-trt1   0.865  0.1738 1.5562 0.0120
```

Each line in the above output contains information about a specific
pairwise comparison. For example, the line trt1-ctrl says that the
comparison of level trt1 with ctrl is not significant (the p-value is
0.39). The confidence interval for the difference $\mu_2 - \mu_1$ is given
by $[-1.06, 0.32]$. Confidence intervals can be visualized by simply
calling plot.

```
plot(TukeyHSD(fit.plant))
```

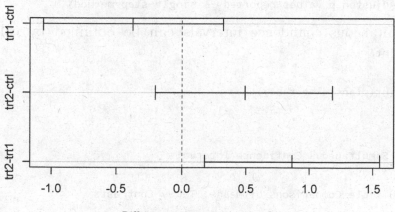

Remember, these confidence intervals are simultaneous, meaning
that the probability that they *all* cover the corresponding true
difference at the same time is 95%. From the p-values, or the
confidence intervals, we read off that only the difference between
trt1 and trt2 is significant (using a significance level of 5%).

We get of course the same results when using package `multcomp`. To do so, we have to use the argument `group = "Tukey"`.

```
## Tukey HSD with package multcomp
plant.glht.tukey <- glht(fit.plant, linfct = mcp(group = "Tukey"))
summary(plant.glht.tukey)
```

```
##
##    Simultaneous Tests for General Linear Hypotheses
##
## Multiple Comparisons of Means: Tukey Contrasts
## ...
## Linear Hypotheses:
##                   Estimate Std. Error t value Pr(>|t|)
## trt1 - ctrl == 0    -0.371      0.279   -1.33    0.391
## trt2 - ctrl == 0     0.494      0.279    1.77    0.198
## trt2 - trt1 == 0     0.865      0.279    3.10    0.012
## (Adjusted p values reported -- single-step method)
```

Simultaneous confidence intervals can be obtained by calling `confint`.

```
confint(plant.glht.tukey)
```

```
##
##    Simultaneous Confidence Intervals
##
## Multiple Comparisons of Means: Tukey Contrasts
## ...
## 95% family-wise confidence level
## ...
## Linear Hypotheses:
##                   Estimate lwr      upr
## trt1 - ctrl == 0 -0.371    -1.062   0.320
## trt2 - ctrl == 0  0.494    -0.197   1.185
## trt2 - trt1 == 0  0.865     0.174   1.556
```

They can be plotted too.

```
plot(confint(plant.glht.tukey))
```

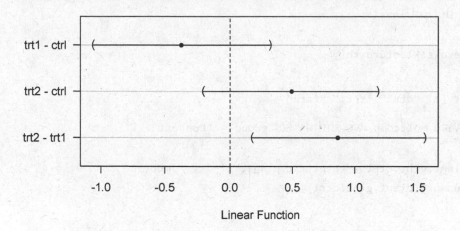

In emmeans, the corresponding function call would be as follows (output not shown):

```
contrast(plant.emm, method = "pairwise")
```

Also with emmeans, the corresponding simultaneous confidence intervals can be obtained with confint, which can be plotted too.

Remark: The implementations in multcomp and emmeans are more flexible with respect to unbalanced data than TukeyHSD, especially for situations where we have multiple factors as for example in Chapter 4.

3.2.5 Multiple Comparisons with a Control

Similarly, if we want to compare all treatment groups with a control group, we have a so-called **multiple comparisons with a control (MCC)** problem (we are basically only considering a *subset* of all pairwise comparisons). The corresponding custom-tailored procedure is called **Dunnett** procedure (Dunnett, 1955). It controls the family-wise error rate in the strong sense and produces simultaneous confidence intervals. As usual, both packages `multcomp` and `emmeans` provide implementations. By default, the first level of the factor is taken as the control group. For the factor `group` in the `PlantGrowth` data set this is `ctrl`, as can be seen when calling the function `levels`.

```
levels(PlantGrowth[,"group"])
```

```
## [1] "ctrl" "trt1" "trt2"
```

With `multcomp`, we simply set `group = "Dunnett"`.

```
plant.glht.ctrl <- glht(fit.plant, linfct = mcp(group = "Dunnett"))
summary(plant.glht.ctrl)
```

```
##
##    Simultaneous Tests for General Linear Hypotheses
##
## Multiple Comparisons of Means: Dunnett Contrasts
## ...
## Linear Hypotheses:
##                    Estimate Std. Error t value Pr(>|t|)
## trt1 - ctrl == 0     -0.371      0.279   -1.33     0.32
## trt2 - ctrl == 0      0.494      0.279    1.77     0.15
## (Adjusted p values reported -- single-step method)
```

We get smaller p-values than with the Tukey HSD procedure because we have to correct for less tests; there are more comparisons between pairs than there are comparisons to the control treatment.

In `emmeans`, the corresponding function call would be as follows (output not shown):

```
contrast(plant.emm, method = "dunnett")
```

The usual approach with `confint` gives the corresponding simultaneous confidence intervals.

3.2.6 FAQ

Should I only do tests like Tukey HSD, etc. if the F-test is significant?

No, the above-mentioned procedures have a *built-in* correction regarding multiple testing and do *not* rely on a significant F-test; one exception is the Scheffé procedure in Section 3.2.3: If the F-test is not significant, you cannot find a significant contrast. In general, conditioning on the F-test leads to a very conservative approach regarding type I error rate. In addition, the conditional coverage rates of, e.g., Tukey HSD confidence intervals can be very low if we only apply them when the F-test is significant, see also Hsu (1996). This means that if researchers would use this recipe of only using Tukey HSD when the F-test is significant and we consider 100 different applications of Tukey HSD, on average it would happen more than 5 times that the simultaneous 95% confidence intervals would not cover all true parameters. Generally speaking, if you apply a statistical test only after a first test was significant, you are typically walking on thin ice: Many properties of the second statistical tests typically change. This problem is also known under the name of **selective inference**, see for example Benjamini et al. (2009).

Is it possible that the F-test is significant but Tukey HSD yields only insignificant pairwise tests? Or the other way round, Tukey HSD yields a significant difference but the F-test is not significant?

Yes, these two tests might give us contradicting results. However, for most situations, this does not happen, see a comparison of the corresponding rejection regions in Hsu (1996).

How can we explain this behavior? This is basically a question of power. For some alternatives, Tukey HSD has more power because it answers a more precise research question, "which pairs of treatments differ?". On the other hand, the F-test is more flexible for situations where the effect is not evident in treatment pairs but in combinations of multiple treatments. Basically, the F-test answers the question, "is a linear contrast of the cell means different from zero?".

We use the following two extreme data sets consisting of three groups having two observations each.

```
x <- factor(rep(c("A", "B", "C"), each = 2))

y1 <- c(0.50, 0.62,
        0.46, 0.63,
        0.95, 0.86)

y2 <- c(0.23, 0.34,
        0.45, 0.55,
        0.55, 0.66)
```

Let us first visualize the first data set:

```
stripchart(y1 ~ x, vertical = TRUE, pch = 1)
```

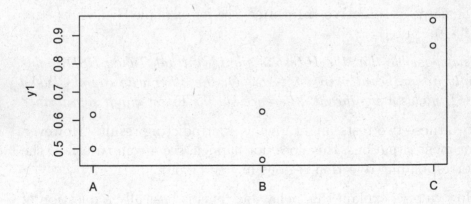

Here, the *F*-test is significant, but Tukey HSD is not:

```
fit1 <- aov(y1 ~ x)
summary(fit1)
```

```
##                 Df Sum Sq Mean Sq F value Pr(>F)
## x               2 0.1659  0.0829    9.68  0.049
## Residuals       3 0.0257  0.0086
```

```
TukeyHSD(fit1)
```

```
##    Tukey multiple comparisons of means
##      95% family-wise confidence level
## ...
##         diff      lwr     upr   p adj
## B-A -0.015 -0.40177  0.3718  0.9857
## C-A  0.345 -0.04177  0.7318  0.0669
## C-B  0.360 -0.02677  0.7468  0.0601
```

Now let us consider the second data set:

```
stripchart(y2 ~ x, vertical = TRUE, pch = 1)
```

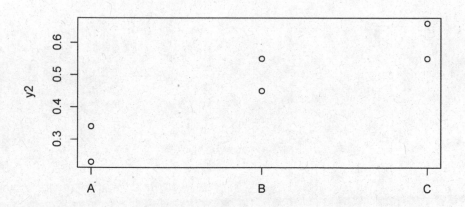

Now, the *F*-test is *not* significant, but Tukey HSD is:

```
fit2 <- aov(y2 ~ x)
summary(fit2)
```

```
##              Df Sum Sq Mean Sq F value Pr(>F)
## x            2 0.1064  0.0532    9.34  0.052
## Residuals   3 0.0171  0.0057
```

```
TukeyHSD(fit2)
```

```
##   Tukey multiple comparisons of means
##     95% family-wise confidence level
## ...
##      diff       lwr    upr  p adj
## B-A 0.215 -0.10049 0.5305 0.1269
## C-A 0.320  0.00451 0.6355 0.0482
## C-B 0.105 -0.21049 0.4205 0.4479
```

4

Factorial Treatment Structure

4.1 Introduction

In the completely randomized designs that we have seen so far, the g different treatments had no special "structure". In practice, treatments are often *combinations* of the levels of two or more factors. Think for example of a plant experiment using combinations of light exposure and fertilizer, with yield as response. We call this a **factorial treatment structure** or a factorial design. If we observe all possible combinations of the levels of two or more factors, we call them **crossed**. An illustration of two crossed factors can be found in Table 4.1.

TABLE 4.1: Example of two crossed factors. With "x" we mean that we have data for the specific combination of exposure level (low / medium / high) and fertilizer brand (brand 1 / brand 2).

Exposure / Fertilizer	Brand 1	Brand 2
low	x	x
medium	x	x
high	x	x

With a factorial treatment structure, we typically have questions about both factors and their possible *interplay*:

- Does the effect of exposure (on the expected value of yield) depend on brand (or vice versa)? In other words, is the exposure effect brand-specific?

If the effects do not depend on the other factor, we could also ask:

- What is the effect of changing exposure from, e.g., low to medium on the expected value of yield?

- What is the effect of changing the brand on the expected value of yield?

Let us have a detailed look at another example: A lab experiment was performed to compare mountain bike tires of two different brands, 1 and 2. To this end, the tires were put on a simulation machine allowing for three different undergrounds (soft, rocky and extreme). Each combination of brand and underground was performed three times (using a new tire each time). The response was the driven kilometers until tread depth was reduced by a pre-defined amount. The experiments have been performed in random order. The data can be found in Table 4.2.

TABLE 4.2: Bike data set.

Brand	Soft	Rocky	Extreme
1	1014, 1062, 1040	884, 929, 893	824, 778, 792
2	1095, 1116, 1127	818, 794, 820	642, 633, 692

Clearly, brand and underground are crossed factors. We have three observations for every combination of the levels of brand and underground.

We first read the data (here, the data is already available as an R-object) and make sure that all categorical variables are correctly encoded as factors.

```
book.url <- "https://stat.ethz.ch/~meier/teaching/book-anova"
bike <- readRDS(url(file.path(book.url, "data/bike.rds")))
str(bike)
```

```
## 'data.frame':    18 obs. of  3 variables:
## $ brand      : Factor w/ 2 levels "1","2": 1 1 1 1 1 1 ...
## $ underground: Factor w/ 3 levels "soft","rocky",..: 1 1 ..
## $ dist       : num  1014 1062 1040 ...
```

With R, we can easily count the observations with the function xtabs.

```
xtabs(~ brand + underground, data = bike)
```

```
##        underground
## brand soft rocky extreme
##     1    3     3       3
##     2    3     3       3
```

The formula "~ brand + underground" in xtabs reads as "count the number of observations for every combination of the levels of the two factors brand and underground."

Typical questions could be as follows:

- Does the effect of brand on the expected value of the distance depend on the underground, or vice versa?

If the effects do not depend on the other factor, we could also ask:

- What is the effect of changing the underground from, e.g., soft to rocky on the expected value of the distance?

- What is the effect of changing the brand from 1 to 2 on the expected value of the distance?

We can visualize such kind of data with a so-called **interaction plot** using the function interaction.plot. For every combination of the levels of brand (1 and 2) and underground (soft, rocky and extreme), we calculate the average value of the response (here, dist). We use underground on the x-axis (the first argument, also called x.factor). In addition, settings corresponding to the same level of brand are connected with lines (argument trace.factor). The role of underground and brand can of course also be interchanged. Hence, what factor we use as x.factor or as trace.factor is a question of personal preference. Most often it is natural to use an ordered factor as the x.factor, as we just did.

```
## elegant way, using the function "with"
with(bike, interaction.plot(x.factor = underground,
                            trace.factor = brand, response = dist))
## standard way: interaction.plot(x.factor = bike$underground,
##                                 trace.factor = bike$brand,
##                                 response = bike$dist)
```

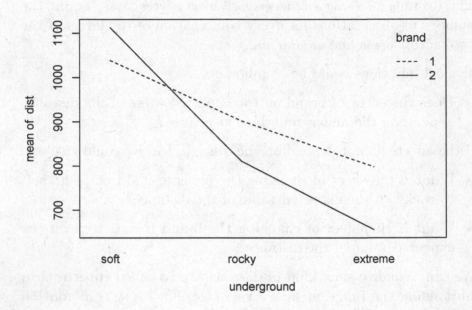

FIGURE 4.1: Interaction plot for the mountain bike tire example.

In Figure 4.1 we can observe that the average value of distance decreases when changing the underground from soft to rocky to extreme. This holds true for both brands. However, the drop of brand 2 when moving from soft to rocky is more pronounced compared to brand 1. One disadvantage of an interaction plot is the fact that it only plots *averages* and we do *not* see the underlying variation of the individual observations anymore. Hence, we cannot easily conclude that the effects which we observe in the plot are statistically significant or not.

A more sophisticated version could be plotted with the package ggplot2 (Wickham, 2016) where we plot both the individual observations and the lines (for interested readers only).

```
library(ggplot2)
ggplot(bike, aes(x = underground, y = dist, lty = brand,
                 shape = brand)) +
  geom_point() + stat_summary(fun = mean, geom = "line",
                 aes(group = brand)) +
  theme_bw()
```

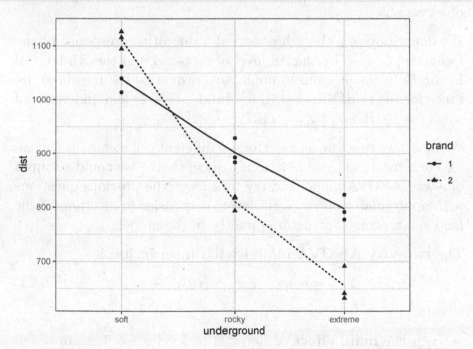

With this plot, we immediately see that the effects are rather large compared to the variation of the individual observations. If we have many observations, we could also use boxplots for every setting instead of the individual observations.

Let us now set up a parametric model which allows us to make a statement about the uncertainty of the estimated effects.

4.2   Two-Way ANOVA Model

We assume a general setup with a factor A with a levels, a factor B with b levels and n replicates for **every** combination of A and B, a **balanced** design. Hence, we have a total of $N = a \cdot b \cdot n$ observations. In the previous example, A was brand with $a = 2$, B was underground with $b = 3$ and we had $n = 3$ replicates for each treatment combination. This resulted in a total of $N = 18$ observations.

We denote by y_{ijk} the kth observed value of the response of the treatment formed by the ith level of factor A and the jth level of factor B. In the previous example, y_{213} would be the measured distance for the third tire ($k = 3$) of brand 2 ($i = 2$) and underground soft ($j = 1$). Hence, $y_{213} = 1127$.

We now have to set up a model for $a \cdot b$ different cell means, or a mean for each treatment combination. Conceptually, we could set up a one-way ANOVA model and try to answer the previous questions with appropriate contrasts. However, it is easier to incorporate the factorial treatment structure directly in the model.

The **two-way ANOVA model with interaction** is

$$Y_{ijk} = \mu + \alpha_i + \beta_j + (\alpha\beta)_{ij} + \epsilon_{ijk} \qquad (4.1)$$

where

- α_i is the **main effect** of factor A at level i, $i = 1, \dots, a$.
- β_j is the **main effect** of factor B at level j, $j = 1, \dots, b$.
- $(\alpha\beta)_{ij}$ is the **interaction effect** between A and B for the level combination i, j (it is *not* the product $\alpha_i\beta_j$).
- ϵ_{ijk} are i.i.d. $N(0, \sigma^2)$ **errors**, $k = 1, \dots, n$.

The expected values for the different cells in the bike tire example can be seen in Table 4.3.

TABLE 4.3: Expected values for the different cells in the bike tire example.

Brand	Soft	Rocky	Extreme
1	$\mu+\alpha_1+\beta_1+(\alpha\beta)_{11}$	$\mu+\alpha_1+\beta_2+(\alpha\beta)_{12}$	$\mu+\alpha_1+\beta_3+(\alpha\beta)_{13}$
2	$\mu+\alpha_2+\beta_1+(\alpha\beta)_{21}$	$\mu+\alpha_2+\beta_2+(\alpha\beta)_{22}$	$\mu+\alpha_2+\beta_3+(\alpha\beta)_{23}$

As usual, we'll have to use a side constraint for the parameters. We will use the sum-to-zero constraint, but we will handle the technical details later (note that the following interpretation is valid under this side constraint).

The effects can be interpreted as follows:

Think of main effects as "average effects" on the expected value of the response when changing the level of a factor. If we consider again Table 4.2 or Table 4.3, the main effects tell us something about average effects when moving from row to row (main effect of brand) and from column to column (main effect of underground).

The interaction effect contains the "leftovers" after having considered the main effects: Are there *cell-specific* properties that cannot be explained well by the main effects? In that sense, the interaction effect measures how far the cell means differ from a main effects only model. Or in other words, the interaction effect captures cell-specific behavior that was not captured with only row and column effects. If the model contains an interaction effect, the effect of, e.g., factor B (underground) depends on the level of factor A (brand). Otherwise, the whole effect could have been captured by the corresponding main effect!

An illustration of the model, for a general setup, can be found in Figure 4.2. This plot is nothing more than a theoretical version of an interaction plot where we have the expected value instead of the sample average on the y-axis. We can observe that the model is flexible enough such that each treatment combination can get its very own mean value ("cell mean").

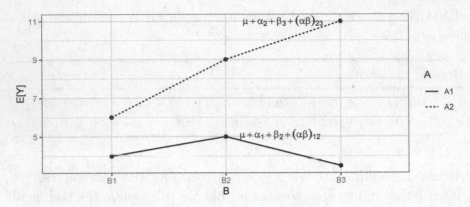

FIGURE 4.2: Illustration of a general factorial model.

A model *without* interaction term is **additive**. This means that the effect of A does not depend on the level of B, and vice versa, "it is always the same, no matter what the level of the other factor". An illustration (for a general setup) can be found in Figure 4.3, where we see that the lines are *parallel*. This means that changing A from level 1 to level 2 has always the same effect (here, an increase by 2 units). Similarly, the effect of changing B does not depend on the level of A (here, an increase by 1 and 3 units if we change B from level 1 to 2, and 2 to 3, respectively).

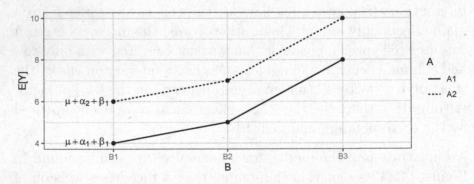

FIGURE 4.3: Illustration of a general factorial model without interaction.

Typically, we use sum-to-zero side constraints, i.e., $\sum_{i=1}^{a} \alpha_i = 0$ and $\sum_{j=1}^{b} \beta_j = 0$ for the main effects. Hence, the main effects have $a - 1$ and $b - 1$ degrees of freedom, respectively, "as before". For the interaction effect we have to make sure that it contains nothing which is row- or column-specific, that is why we have the side constraints $\sum_{i=1}^{a} (\alpha\beta)_{ij} = 0$ (sum across rows) for all j's and $\sum_{j=1}^{b} (\alpha\beta)_{ij} = 0$ (sum across columns) for all i's. Therefore, the degrees of freedom of the interaction term are $(a-1)(b-1)$. Rule to remember: The degrees of freedom of the interaction effect are the product of the degrees of freedom of the involved main effects.

4.2.1 Parameter Estimation

As usual, we estimate parameters using the principles of **least squares**. Using sum-to-zero side constraints, one can show that we get the parameter estimates listed in Table 4.4. As before, if we replace an index with a dot, we take the mean, or the sum, over that dimension. For example, $\overline{y}_{1..}$ is the mean over all observations in the first row: y_{1jk}, $j = 1, \ldots, b$ and $k = 1, \ldots, n$. To calculate $\hat{\alpha}_1$, we simply determine the deviation of the mean of the first row $\overline{y}_{1..}$ from the overall mean $\overline{y}_{...}$, i.e.,

$$\hat{\alpha}_1 = \overline{y}_{1..} - \overline{y}_{...}.$$

TABLE 4.4: Parameter estimates for the two-way ANOVA model.

Parameter	Estimate
μ	$\hat{\mu} = \overline{y}_{...}$
α_i	$\hat{\alpha}_i = \overline{y}_{i..} - \overline{y}_{...}$
β_j	$\hat{\beta}_j = \overline{y}_{.j.} - \overline{y}_{...}$
$(\alpha\beta)_{ij}$	$\widehat{(\alpha\beta)}_{ij} = \overline{y}_{ij.} - \hat{\mu} - \hat{\alpha}_i - \hat{\beta}_j$

Combining all this information, we estimate the expected value of the response Y_{ijk} for A at level i and B at level j as

$$\hat{\mu} + \hat{\alpha}_i + \hat{\beta}_j + \widehat{(\alpha\beta)}_{ij} = \overline{y}_{ij.}.$$

which is nothing more than the mean of the observations in the corresponding cell (which is hopefully no surprise). However, note that we untangled the effect with respect to the two main effects and the interaction. If we would use an unstructured cell means model with $a \cdot b$ different cell means, we would get the very same estimates for the cell means.

If we carefully inspect the parameter estimates in Table 4.4, we observe that for the main effects, we use an estimate that completely ignores the other factor. We basically treat the problem as a one-way ANOVA model. This is a consequence of the balanced design. For all levels of A we have the "same population" of levels of B. Hence, if we compare $\bar{y}_{1..}$ (average over the first row) with $\bar{y}_{2..}$ (average over second row), the effect is only due to changing A from level 1 to level 2. In regression terminology, we would call this an orthogonal design.

In R, we fit the model using the function aov. For our example, the model of Equation (4.1) would be written as dist ~ brand + underground + brand:underground in R formula notation. The colon ":" indicates an interaction effect. Separate terms are connected by a "+". Alternatively, the asterisk "*" is a useful shorthand notation to include two main effects as well as their interaction, i.e., the same model could also be specified as dist ~ brand * underground. A model which contains only the main effects of brand and underground (without interaction effect) would be specified as dist ~ brand + underground.

```
options(contrasts = c("contr.sum", "contr.poly"))
fit.bike <- aov(dist ~ brand * underground, data = bike)
coef(fit.bike)
```

```
##        (Intercept)                brand1         underground1
##             886.28                 26.61               189.39
##       underground2  brand1:underground1  brand1:underground2
##             -29.94                -63.61                19.06
```

The remaining coefficients can be obtained by making use of the sum-to-zero constraint. An easier way is to use the function dummy.coef

```
dummy.coef(fit.bike)
```

```
## Full coefficients are
##
## (Intercept):          886.3
## brand:                  1       2
##                       26.61  -26.61
## underground:          soft    rocky extreme
##                      189.39  -29.94 -159.44
## brand:underground:   1:soft  2:soft 1:rocky 2:rocky
##                      -63.61   63.61   19.06  -19.06
## ...
## brand:underground:   1:extreme 2:extreme
##                         44.56    -44.56
```

For example, the cell mean of combination brand 1 and underground soft can be explained as the sum of (Intercept): $\hat{\mu} = 886.28$, main effect of brand 1: $\hat{\alpha}_1 = 26.61$, main effect of underground soft: $\hat{\beta}_1 = 189.39$ and interaction effect of the *combination* of the two levels of brand and underground 1:soft: $\widehat{(\alpha\beta)}_{11} = -63.61$.

As an exercise and for a better understanding of the parameter estimates, you can of course also manually calculate, or verify, the parameter estimates using the information in Table 4.5.

TABLE 4.5: Bike data set including row-, column-wise and overall mean.

Brand	Soft	Rocky	Extreme	$\bar{y}_{i..}$
1	1014, 1062, 1040	884, 929, 893	824, 778, 792	$\bar{y}_{1..} = 912.89$
2	1095, 1116, 1127	818, 794, 820	642, 633, 692	$\bar{y}_{2..} = 859.67$
$\bar{y}_{.j.}$	$\bar{y}_{.1.} = 1075.67$	$\bar{y}_{.2.} = 856.33$	$\bar{y}_{.3.} = 726.83$	$\bar{y}_{...} = 886.28$

4.2.2   Tests

As in the case of the one-way ANOVA, the total sum of squares SS_T can be (uniquely) **partitioned** into **different sources**, i.e.,

$$SS_T = SS_A + SS_B + SS_{AB} + SS_E,$$

where the individual terms are given in Table 4.6.

TABLE 4.6: Sum of squares for the two-way ANOVA model.

Source	Sum of Squares
A ("between rows")	$SS_A = \sum_{i=1}^{a} bn(\hat{\alpha}_i)^2$
B ("between columns")	$SS_B = \sum_{j=1}^{b} an(\hat{\beta}_j)^2$
AB ("correction")	$SS_{AB} = \sum_{i=1}^{a} \sum_{j=1}^{b} n(\widehat{\alpha\beta})_{ij}^2$
Error ("within cells")	$SS_E = \sum_{i=1}^{a} \sum_{j=1}^{b} \sum_{k=1}^{n} (y_{ijk} - \overline{y}_{ij.})^2$
Total	$SS_T = \sum_{i=1}^{a} \sum_{j=1}^{b} \sum_{k=1}^{n} (y_{ijk} - \overline{y}_{...})^2$

By calculating the corresponding mean squares, we can construct an ANOVA table, see Table 4.7.

TABLE 4.7: ANOVA table for the two-way ANOVA model.

Source	df	SS	Mean Squares	F-ratio
A	$a-1$	SS_A	$MS_A = \frac{SS_A}{a-1}$	$\frac{MS_A}{MS_E}$
B	$b-1$	SS_B	$MS_B = \frac{SS_B}{b-1}$	$\frac{MS_B}{MS_E}$
AB	$(a-1)(b-1)$	SS_{AB}	$MS_{AB} = \frac{SS_{AB}}{(a-1)(b-1)}$	$\frac{MS_{AB}}{MS_E}$
Error	$ab(n-1)$	SS_E	$MS_E = \frac{SS_E}{ab(n-1)}$	

Note: The degrees of freedom of the error term is nothing more than the degrees of freedom of the total sum of squares ($abn - 1$, number of observations minus 1) minus the sum of the degrees of freedom of all other effects.

As in the one-way ANOVA situation, we can construct global tests for the main effects and the interaction effect:

Interaction effect: The null hypothesis that there is no interaction effect can be translated as: "The effect of factor A does not depend on the level of factor B (or the other way round). Or, the main effects model is enough." In the example it would mean: "The effect of underground does not depend on brand." The model could be visualized as in Figure 4.3, i.e., with *parallel* lines. In formulas we have

$$H_0 : (\alpha\beta)_{ij} = 0 \text{ for all } i, j$$
$$H_A : \text{at least one } (\alpha\beta)_{ij} \neq 0$$

Under H_0 it holds that $\frac{MS_{AB}}{MS_E} \sim F_{(a-1)(b-1),\, ab(n-1)}$.

Main effect of A: The null hypothesis that there is no main effect of A can be translated as: "The (average) effect of factor A does not exist." In the example it would mean: "There is no effect of brand when averaged over all levels of underground. Or, both brands have the same expected distance when averaged over all levels of underground." In formulas we have

$$H_0 : \alpha_i = 0 \text{ for all } i$$
$$H_A : \text{at least one } \alpha_i \neq 0$$

Under H_0 it holds that $\frac{MS_A}{MS_E} \sim F_{a-1,\, ab(n-1)}$.

Main effect of B: The null hypothesis that there is no main effect of B can be translated as: "The (average) effect of factor B does not exist." In the example it would mean: "There is no effect of underground when averaged over both brands. Or, all undergrounds have the same expected distance when averaged over both brands." In formulas we have

$$H_0 : \beta_j = 0 \text{ for all } j$$
$$H_A : \text{at least one } \beta_j \neq 0$$

Under H_0 it holds that $\frac{MS_B}{MS_E} \sim F_{b-1,\, ab(n-1)}$.

The *F*-distributions all follow the same pattern: The degrees of freedom are given by the numerator and the denominator of the corresponding *F*-ratio, respectively.

In R, we get the ANOVA table and the corresponding p-values again with the summary function. We read the following R output from *bottom* to *top*. This means that we first check whether we need the interaction term or not. If there is no evidence of interaction, we continue with the inspection of the main effects.

```
summary(fit.bike)
```

```
##                   Df Sum Sq Mean Sq F value  Pr(>F)
## brand              1  12747   12747    24.0 0.00037
## underground        2 373124  186562   351.5 2.2e-11
## brand:underground  2  38368   19184    36.1 8.3e-06
## Residuals         12   6369     531
```

There is evidence of an interaction between brand and underground as the p-value is very small (line brand:underground). This finding is consistent with what we have observed in the interaction plot in Figure 4.1. This means that there is statistical evidence that the effect of underground depends on brand. Typically, the effect of underground is then analyzed for each level of brand separately. From the interaction plot we can observe that this interaction is mainly caused by the different behavior for underground soft. In the presence of a significant interaction, the main effects (which are average effects) are typically not very interesting anymore. For example, if we have a data set where the interaction is significant but one of the main effects is not, we should *not* conclude that there is no effect of the corresponding factor. The effect simply depends on the level of the other factor. Some of the possible situations for the true parameter values of Equation (4.1) for the situation of two factors with two levels each are illustrated in Figure 4.4. In panel 1, the strength of the effect of A depends on the level of B, but the sign is always the same (both the main effects of A and B and the interaction are nonzero). In panel 2, there is no effect of A for B at level 1, but at level 2 (again, both the main effects of A and B and the interaction are nonzero). In panel 3 we have the very special situation that both main effects vanish, but the interaction is very pronounced. In panel 4, the sign of the effect of A depends on the level of B (both the main effects of A and B and the interaction are nonzero).

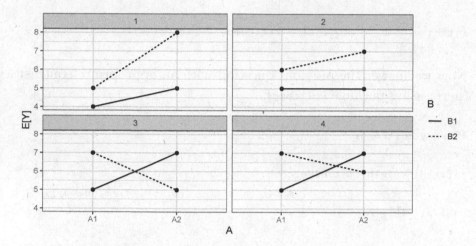

FIGURE 4.4: Illustration of different parameter settings for two factors having two levels each.

In our example, we could start analyzing the effect of brand for each level of underground by inspecting some specific questions using appropriate contrasts (this is of course also meaningful if such a question was planned beforehand). For example, what is the difference between the two brands for underground soft? From a technical point of view, one way of doing this is by constructing what we would call a **hyper-factor** which has as levels *all* possible combinations of brand and underground; this is equivalent to the full model but typically easier to handle with respect to the mentioned contrast. In R, we can easily do this with the function interaction which "glues together" the information from two factors.

```
bike[,"trt.comb"] <- interaction(bike[, "brand"],
                          bike[, "underground"])
levels(bike[,"trt.comb"])
```

```
## [1] "1.soft"    "2.soft"    "1.rocky"   "2.rocky"
## [5] "1.extreme" "2.extreme"
```

We use this hyper-factor to fit a one-way ANOVA model. Again, this gives the same estimated cell means as the two-way ANOVA model, but it is easier for handling the contrast later on.

```
fit.bike.comb <- aov(dist ~ trt.comb, data = bike)
```

Now we answer the previous question with an appropriate contrast (here, using package multcomp).

```
library(multcomp)
## Brand 1 vs. brand 2 for underground soft
fit.glht <- glht(fit.bike.comb,
                 linfct = mcp(trt.comb = c(1, -1, 0, 0, 0, 0)))
confint(fit.glht)
```

```
## ...
## Linear Hypotheses:
##          Estimate lwr      upr
## 1 == 0   -74.000 -114.983  -33.017
```

This means that for underground soft, a 95% confidence interval for the difference between brand 1 and brand 2 is given by $[-114.98, -33.02]$.

Note: Both multcomp and emmeans can be used to directly handle contrasts using the two-way ANOVA model (with interaction). However, interpretation can be difficult (typically you would also see a corresponding warning message). This is why we prefer to explicitly fit the corresponding cell means model and apply the contrast there.

Whether an interaction is needed or not might also depend on the *scale* on which we are analyzing the response. A famous example is the logarithm. Effects that are *multiplicative* on the original scale become *additive* on the log-scale, i.e., no interaction is needed on the log-scale. For example, see Figure 4.5. In the left panel, we can observe that there might be an interaction effect and that variation is larger for large values of the response. In the right panel, we see the corresponding interaction plot when taking the logarithm of the response. The interaction effect seems to have gone and in addition, variability is much more constant than before. This phenomenon is very common in practice.

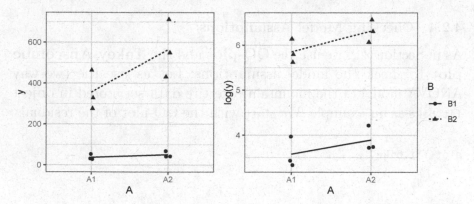

FIGURE 4.5: Interaction plot on the original scale (left) and on the log-scale (right) for some simulated data. Some (horizontal) jittering was applied to avoid overplotting of points.

4.2.3 Single Observations per Cell

If we only have a *single* observation in each "cell" ($n = 1$), we *cannot* do statistical inference anymore with a model including the interaction. The reason is that we have no idea of the experimental error (the variation between experimental units getting the *same* treatment). From a technical point of view, fitting a model with an interaction term would lead to a perfect fit with all residuals being exactly zero.

However, we can still fit a main effects only model. If the data generating mechanism actually contains an interaction, we are fitting a *wrong* model. The consequence is that the estimate of the error variance will be biased (upward). Hence, the corresponding tests will be too conservative, meaning p-values will be too large and confidence intervals too wide. This is not a problem as the type I error rate is still controlled; we "just" lose power.

An alternative approach would be to use a Tukey one-degree of freedom interaction (Tukey, 1949b), a highly structured form of interaction effect where the interaction effect actually is assumed to be proportional to the product of the corresponding main effects. This would only use _one_ additional parameter for the interaction term.

4.2.4 Checking Model Assumptions

As in Section 2.2, we use the **QQ-plot** and the **Tukey-Anscombe plot** to check the model assumptions. Let us use the two-way ANOVA model of the mountain bike tire data set stored in object `fit.bike` as an example. We start with the QQ-plot of the residuals:

```
plot(fit.bike, which = 2)
```

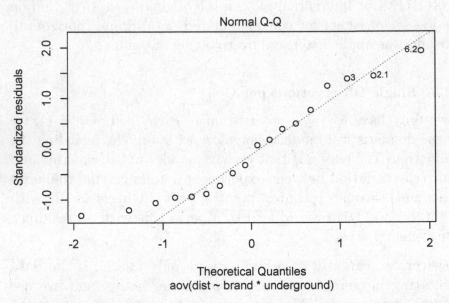

FIGURE 4.6: QQ-plot of the residuals of the two-way ANOVA model of the mountain bike tire data set.

If in doubt that the QQ-plot in Figure 4.6 is still OK, we could use the function qqPlot of the package car as discussed in Section 2.2.1. Alternatively, we can always compare it to a couple of "nice" simulated QQ-plots of the same sample size using the following command (give it a try, output not shown).

```
qqnorm(rnorm(nrow(bike)))
```

Our QQ-plot in Figure 4.6 looks like one from simulated data. Hence, there is no evidence of a violation of the normality assumption of the error term.

The Tukey-Anscombe plot looks OK too in the sense that there is no evidence of non-constant variance:

```
plot(fit.bike, which = 1, add.smooth = FALSE)
```

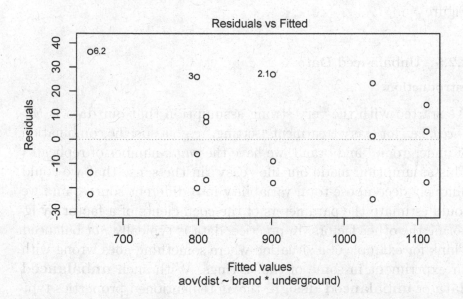

To illustrate, the Tukey-Anscombe plots for two-way ANOVA models (including interaction) using the data of Figure 4.5 are shown in Figure 4.7. On the original scale, a funnel-like shape is clearly visible. This is a sign that transforming the response with the logarithm will stabilize the variance, as can be observed from the plot on the transformed data.

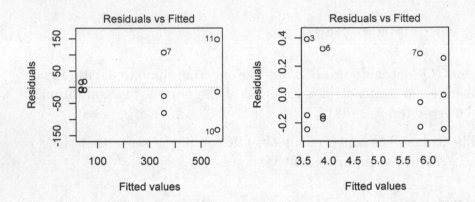

FIGURE 4.7: Tukey-Anscombe plot of the model on the original scale (left) and the model on the log-scale (right) for the data of Figure 4.5.

4.2.5 Unbalanced Data

Introduction

We started with the very strong assumption that our data is *balanced*, i.e., for every treatment "setting," e.g., a specific combination of underground and brand, we have the *same* number of replicates. This assumption made our life "easy" in the sense that we could uniquely decompose total variability into different sources and we could estimate the parameters of the coefficients of a factor by ignoring the other factors. In practice, data is typically *not* balanced. Think for example of a situation where something goes wrong with an experiment in some of the settings. With such **unbalanced data** or **unbalanced design**, the aforementioned properties typically do not hold anymore. We still use least squares to estimate the parameters. The estimates will look more complicated. However, this is *not* a problem these days as we can easily calculate them using R. More of a problem is the fact that the sum of squares *cannot* be *uniquely* partitioned into different sources anymore. This means that for some part of the variation of the data, it is *not* clear what source we should attribute it to.

Typically, people use an appropriate **model comparison** approach instead. As seen before, for example in Table 4.6, the sum of squares of a factor can be interpreted as the **reduction of residual sum of squares** when adding the corresponding factor to the model. In the balanced situation, it would not matter whether the remaining factors are in the model or not; the reduction is always the same. Unfortunately, for unbalanced data this property does *not* hold anymore.

We follow Oehlert (2000) and use the following notation: With SS(B | 1, A) we denote the **reduction in residual sum of squares** when comparing the model (1, A, B) (= y ~ A + B) with (1, A) (= y ~ A). The 1 is the overall mean μ (that does not appear explicitly in the R model formula). Interpretation of the corresponding test is as follows: "Do we need factor B in the model if we already have factor A, or after having controlled for factor A?". There are different "ways" or "types" of model comparison approaches:

- Type I (sequential): Sequentially build up model (depends on the ordering of the model terms!)
    - SS(A | 1)
    - SS(B | 1, A)
    - SS(AB | 1, A, B)
- Type II (hierarchical): Controlling for the influence of the largest *hierarchical* model not including the term of interest. With hierarchical we mean that whenever an interaction is included in the model, the corresponding main effects are included too.
    - SS(A | 1, B)
    - SS(B | 1, A)
    - SS(AB | 1, A, B)
- Type III (fully adjusted): Controlling for *all* other terms.
    - SS(A | 1, B, AB)
    - SS(B | 1, A, AB)
    - SS(AB | 1, A, B)

Type I is what you get with summary of an aov object (like summary(fit.bike)) or if you call the function anova using a single argument, e.g., anova(fit.bike). Hence, for unbalanced data you get different results whether you write y ~ A * B or y ~ B * A, see also R FAQ 7.18[1]. For type II we can either use the function Anova in the package car or we could compare the appropriate models with the function anova ourselves. For type III we can use the command drop1; we have to be careful that we set the contrast option to contr.sum in this special situation for technical reasons, see also the warning in the help file of the function Anova of package car.

Typically, we take MS_E from the *full* model (including all terms) as the estimate for the error variance to construct the corresponding F-tests.

Note that interpretation of the test depends on the type that we use. This can be seen from the following remarks.

Some Technical Remarks

We use the notation
$$Y_{ijk} = \mu_{ij} + \epsilon_{ijk}$$
where $\mu_{ij} = \mu + \alpha_i + \beta_j + (\alpha\beta)_{ij}$. This means we use a cell means model with expected values μ_{ij}. We denote the corresponding sample sizes for the level combination (i, j) with $n_{ij} > 0$. It can be shown (Speed et al., 1978) that both type I and type II sum of squares lead to tests that involve null hypotheses that depend on the sample sizes of the individual cells!

More precisely, for example for the main effect A we have:

- For type I sum of squares, we implicitly test the null hypothesis

$$H_0 : \sum_{j=1}^{b} \frac{n_{1j}}{n_{1\cdot}} \mu_{1j} = \cdots = \sum_{j=1}^{b} \frac{n_{aj}}{n_{a\cdot}} \mu_{aj}$$

where $n_{i\cdot}$ is the sum of the corresponding n_{ij}'s. The above equation simply means that across all rows, a weighted mean

---

[1] https://cran.r-project.org/doc/FAQ/R-FAQ.html#Why-does-the-output-from-anova_0028_0029-depend-on-the-order-of-factors-in-the-model_003f

of the μ_{ij}'s is the same. Note that the weights depend on the *observed* sample sizes n_{ij}. Therefore, the observed sample sizes dictate our null hypothesis! Interpretation of such a research question is rather difficult.

- Similarly, for type II sum of squares we get a more complicated formula involving n_{ij}'s.

- For type III sum of squares we have

$$H_0 : \overline{\mu}_{1\cdot} = \cdots = \overline{\mu}_{a\cdot}$$

where $\overline{\mu}_{i\cdot} = \frac{1}{b}\sum_{l=1}^{b}\mu_{il}$ which is the *unweighted* mean of the corresponding μ_{ij}'s. This can be reformulated as

$$H_0 : \alpha_1 = \cdots = \alpha_a$$

This is what we have used so far, and more importantly, does *not* depend on n_{ij}.

We can also observe that for a *balanced* design, we test the same null hypothesis, no matter what type we use.

Unbalanced Data Example

We have a look at some simulated data about a sports experiment using a factorial design with the factor "training method" (method, with levels conventional and new) and the factor "energy drink" (drink, with levels A and B). The response is running time in seconds for a specific track.

```
book.url <- "https://stat.ethz.ch/~meier/teaching/book-anova"
running <- read.table(file.path(book.url, "data/running.dat"),
                      header = TRUE)
str(running)
```

```
## 'data.frame':    70 obs. of  3 variables:
##  $ method: chr  "new" "new" ...
##  $ drink : chr  "A" "A" ...
##  $ y     : num  40.6 49.7 42.1 42.2 39 44.2 ...
```

The output of xtabs gives us all the n_{ij}'s.

```
xtabs(~ method + drink, data = running)
```

```
##                   drink
## method           A  B
##    conventional 10 40
##    new          10 10
```

Clearly, this is an unbalanced data set. We use contr.sum; otherwise, type III sum of squares will be wrong (technical issue).

```
options(contrasts = c("contr.sum", "contr.poly"))
## Type I sum of squares
fit <- aov(y ~ method * drink, data = running)
summary(fit)
```

```
##               Df Sum Sq Mean Sq F value  Pr(>F)
## method         1   2024    2024  263.72 < 2e-16
## drink          1    455     455   59.32 9.1e-11
## method:drink   1     29      29    3.79   0.056
## Residuals     66    507       8
```

For example, the row method tests the null hypothesis

$$H_0 : 0.2 \cdot \mu_{11} + 0.8 \cdot \mu_{12} = 0.5 \cdot \mu_{21} + 0.5 \cdot \mu_{22}$$

(because 80% of training method conventional use energy drink B).

Now we change the order of the factors in the model formula.

```
fit2 <- aov(y ~ drink * method, data = running)
summary(fit2) ## sum of squares of method change!
```

```
##               Df Sum Sq Mean Sq F value Pr(>F)
## drink          1   1146    1146  149.30 <2e-16
## method         1   1333    1333  173.74 <2e-16
## drink:method   1     29      29    3.79  0.056
## Residuals     66    507       8
```

We can see that the sum of squares depend on the ordering of the model terms in the model formula if we use a type I approach (if `method` comes first, we get 2024, otherwise 1333). Hence, we also get different F-ratios and different p-values. However, the p-values of the main effects are very small here, no matter what order we use.

We can easily get type II sum of squares with the package `car`.

```
## Type II sum of squares
library(car)
Anova(fit, type = "II", data = running)
```

```
## Anova Table (Type II tests)
##
## Response: y
##             Sum Sq Df F value  Pr(>F)
## method        1333  1  173.74 < 2e-16
## drink          455  1   59.32 9.1e-11
## method:drink    29  1    3.79   0.056
## Residuals      507 66
```

We could also use the `Anova` function for type III sum of squares, but `drop1` will do the job too.

```
## Type III sum of squares
drop1(fit, scope = ~., test = "F", data = running)
## or: Anova(fit, type = "III", data = running)
```

```
## Single term deletions
##
## Model:
## y ~ method * drink
##              Df Sum of Sq  RSS AIC F value  Pr(>F)
## <none>                     507 146
## method        1      1352 1859 236  176.21 < 2e-16
## drink         1       484  991 192   63.09 3.3e-11
## method:drink  1        29  536 148    3.79   0.056
```

Now the row method tests the null hypothesis

$$H_0 : \frac{1}{2} \cdot \mu_{11} + \frac{1}{2} \cdot \mu_{12} = \frac{1}{2} \cdot \mu_{21} + \frac{1}{2} \cdot \mu_{22},$$

which is nothing but saying that the (unweighted) "row-average" of the μ_{ij}'s is constant. Here, the actual sample sizes do not play a role anymore. Note that we always get the same result for the interaction effect, no matter what type we use.

4.3 Outlook

4.3.1 More Than Two Factors

We can easily extend the model to more than two factors. If we have three factors A, B and C (with a, b and c levels, respectively), we have 3 main effects, $3 \cdot 2/2 = 3$ two-way interactions (the "usual" interaction so far) and one so-called **three-way interaction**. We omit the mathematical model formulation and work directly with the corresponding R code. In R, we would write y ~ A * B * C for a **three-way ANOVA model** including all main effects, all two-way interactions and a three-way interaction. An equivalent formulation would be y ~ A + B + C + A:B + A:C + B:C + A:B:C.

Main effects are interpreted as average effects, two-way interaction effects are interpreted as deviations from the main effects model, i.e., the correction for an effect that depends on the level of the other factor, and the three-way interaction is an adjustment of the two-way interaction depending on the third factor. Or in other words, if there is a three-way interaction it means that the effect of factor A depends on the level combination of the factors B and C, i.e., each level combination of B and C has its own effect of A. This typically makes interpretation difficult.

In a balanced design, the model parameters are estimated as "usual." To estimate the three-way interaction, we simply subtract all estimates of the main effects and the two-way interactions. The degrees of freedom for the three-way interaction are the product of the degrees of freedom of all involved factors.

A potential ANOVA table for a balanced design with n replicates in each cell is given in Table 4.8.

TABLE 4.8: ANOVA table for the three-way ANOVA model.

Source	df	F-ratio
A	$a-1$	$\frac{MS_A}{MS_E}$
B	$b-1$	$\frac{MS_B}{MS_E}$
C	$c-1$	$\frac{MS_C}{MS_E}$
AB	$(a-1)(b-1)$	$\frac{MS_{AB}}{MS_E}$
AC	$(a-1)(c-1)$	$\frac{MS_{AC}}{MS_E}$
BC	$(b-1)(c-1)$	$\frac{MS_{BC}}{MS_E}$
ABC	$(a-1)(b-1)(c-1)$	$\frac{MS_{ABC}}{MS_E}$
Error	$abc(n-1)$	

The model with a three-way interaction is flexible enough to model abc different cell means! We also observe that the three-way interaction typically has large degrees of freedom (meaning, it needs a lot of parameters!). We can only do statistical inference about the three-way interaction if we have multiple observations in the individual cells. If we would only have one observation in each cell ($n = 1$), we could still fit the model y ~ A + B + C + A:B + A:C + B:C. This means that we drop the highest-order interaction and "pool it" into the error term. This is quite a common strategy, especially if we have more than three factors. Most often, the effect size of the most complex interaction is assumed to be small or zero. Hence, the corresponding term can be dropped from the model to save degrees of freedom. Ideally, these decisions are made *before* looking at the data. Otherwise, dropping all the insignificant terms from the model and putting them into the error term will typically lead to *biased* results (see also the comments in section 8.9 in Oehlert, 2000). This will make the remaining model terms look too significant, i.e., we declare too many effects as significant although they are actually not. This means that the type I error rate is *not* being controlled anymore. In addition, the corresponding confidence intervals will be too narrow.

A common question is: "Can I test a certain interaction with my data?". If multiple observations are available for each combination of the levels of the involved factors, we can test the interaction. We do *not* have to consider the remaining factors in the model. For example, if we have three factors A, B and C, we can do statistical inference about the interaction between A and B if we have multiple observations for each combination of the levels of A and B, even if they correspond to *different* levels of C. Hence, as already mentioned, even if we have only one observation for each treatment combination of A, B and C we can do statistical inference about two-way interaction effects (but *not* about the three-way interaction).

4.3.2 Nonparametric Alternatives

Unfortunately, nonparametric approaches are not very common for factorial designs. For example, when using randomization tests, it is much more difficult to isolate the effect of interest (say, a certain main effect). More information can be found for example in Edgington and Onghena (2007) or Anderson (2001).

5

Complete Block Designs

5.1 Introduction

In many situations we know that our experimental units are *not* homogeneous. Making explicit use of the special structure of the experimental units typically helps reduce variance ("getting a more precise picture"). In your introductory course, you have learned how to apply the paired t-test. It was used for situations where two treatments were applied on the same "object" or "subject." Think for example of applying two treatments, in parallel, on human beings (like the application of two different eye drop types, each applied in one of the two eyes). We know that individuals can be very different. Due to the fact that we apply both treatments on the same subject, we get a "clear picture" of the treatment effect *within* every subject by taking the difference of the response values corresponding to the two treatments. This makes the subject-to-subject variability completely disappear. We also say that we block on subjects or that an individual subject is a block. This is also illustrated in Figure 5.1 (left): The values of the same subject are connected with a line. With the help of these lines, it is obvious that the response value corresponding to treatment is larger than the value corresponding to the control group, within a subject. This would be much less obvious if the data would come from two independent groups. In this scenario, we would have to delete the lines and would be confronted with the *whole* variability *between* the different subjects.

We will now extend this way of thinking to the situation $g > 2$, where g is the number of levels of our treatment factor (as in Chapter 2). An illustration of the basic idea can be found in

Figure 5.1 (right). We simply consider situations where we have more than two levels on the x-axis. The basic idea stays the same: Values coming from the same block (here, subject) can be connected with a line. This helps both our eyes and the statistical procedure in getting a much clearer picture of the treatment effect.

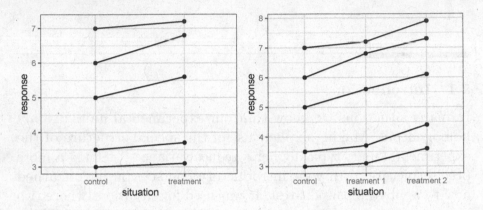

FIGURE 5.1: Illustration of the basic idea of blocking. Left: Situation of paired samples, values coming from the same subject are connected with a line. Right: Extension to the situation where we apply more than two treatments on the same subject.

5.2   Randomized Complete Block Designs

Assume that we can divide our experimental units into r groups, also known as **blocks**, containing g experimental units each. Think for example of an agricultural experiment at r different locations having g different plots of land each. Hence, a block is given by a location and an experimental unit by a plot of land. In the introductory example, a block was given by an individual subject.

The **randomized complete block design (RCBD)** uses a **restricted randomization scheme**: *Within* every block, e.g., at each location, the g treatments are randomized to the g experimental units, e.g., plots of land. In that context, location is also called the **block factor**. The design is called *complete* because we observe the complete set of treatments within every block (we will later also learn about *incomplete* block designs where this is not the case anymore, see Chapter 8).

Note that blocking is a special way to *design* an experiment, or a special "flavor" of randomization. It is *not* something that you use only when analyzing the data. Blocking can also be understood as replicating an experiment on multiple sets, e.g., different locations, of homogeneous experimental units, e.g., plots of land at an individual location. The experimental units should be as similar as possible *within* the same block, but can be very different *between* different blocks. This design allows us to fully remove the between-block variability, e.g., variability between different locations, from the response because it can be explained by the block factor. Hence, we get a much clearer picture for the treatment factor. The randomization step *within* each block makes sure that we are protected from unknown confounding variables. A completely randomized design (ignoring the blocking structure) would typically be much less efficient as the data would be noisier, meaning that the error variance would be larger. In that sense, blocking is a so-called variance reduction technique.

Typical block factors are location (see example above), day (if an experiment is run on multiple days), machine operator (if different operators are needed for the experiment), subjects, etc.

Blocking is very powerful and the general rule is, according to George Box (Box et al., 1978):

Block what you can; randomize what you cannot.

In the most basic form, we assume that we do **not** have replicates within a block. This means that we only observe every treatment *once* in each block.

TABLE 5.1: Conceptual layout of the design for the oat example. Different blocks are denoted by different columns. The three rows correspond to the different plots of land at each location.

1	2	3	4	5	6
Marvellous	Victory	Golden.rain	Marvellous	Marvellous	Golden.rain
Victory	Golden.rain	Victory	Victory	Golden.rain	Marvellous
Golden.rain	Marvellous	Marvellous	Golden.rain	Victory	Victory

The analysis of a randomized complete block design is straightforward. We treat the block factor as "just another" factor in our model. As we have no replicates within blocks, we can only fit a main effects model of the form

$$Y_{ij} = \mu + \alpha_i + \beta_j + \epsilon_{ij},$$

where the α_i's are the treatment effects and the β_j's are the **block effects** with the usual side constraints. In addition, we have the usual assumptions on the error term ϵ_{ij}. According to this model, we implicitly assume that blocks only cause *additive* shifts. Or in other words, the treatment effects are always the same, no matter what block we consider. This assumption is usually made based on domain knowledge. If there would be an interaction effect between the block and the treatment factor, the result would be very difficult to interpret. Most often, this is due to unmeasured variables, e.g., different soil properties at different locations.

We consider an example which is adapted from Venables and Ripley (2002), the original source is Yates (1935) (we will see the full data set in Section 7.3). At six different locations (factor block), three plots of land were available. Three varieties of oat (factor variety with levels Golden.rain, Marvellous and Victory) were randomized to them, individually per location. The response was yield (in 0.25lbs per plot). A conceptual layout of the design can be found in Table 5.1.

The data can be read as follows:

```
block    <- factor(rep(1:6, times = 3))
variety <- rep(c("Golden.rain", "Marvellous", "Victory"), each = 6)
yield    <- c(133.25, 113.25, 86.75, 108, 95.5, 90.25,
                129.75, 121.25, 118.5, 95, 85.25, 109,
                143, 87.25,  82.5, 91.5, 92, 89.5)
oat.variety <- data.frame(block, variety, yield)
xtabs(~ block + variety, data = oat.variety)
```

```
##         variety
## block Golden.rain Marvellous Victory
##    1            1          1       1
##    2            1          1       1
##    3            1          1       1
##    4            1          1       1
##    5            1          1       1
##    6            1          1       1
```

We use the usual `aov` function with a model including the two main effects `block` and `variety`. It is good practice to write the block factor first; in case of unbalanced data, we would get the effect of variety adjusted for block in the sequential type I output of `summary`, see Section 4.2.5 and also Chapter 8.

```
fit.oat.variety <- aov(yield ~ block + variety, data = oat.variety)
summary(fit.oat.variety)
```

```
##              Df Sum Sq Mean Sq F value Pr(>F)
## block         5   3969     794    5.28  0.012
## variety       2    447     223    1.49  0.272
## Residuals    10   1503     150
```

We first focus on the p-value of the treatment factor `variety`. Although we used a randomized complete block design, we cannot reject the null hypothesis that there is no overall effect of variety (a reason might be low power, as we only have 10 degrees of freedom left for the error term). Typically, we are not inspecting the p-value of the block factor `block`. There is some historic debate why we should not do this, mainly because of the fact that we did *not* randomize blocks to experimental units. In addition, we already knew (or hoped) beforehand that blocks are different. Hence, such a finding would not be of great scientific relevance. However, we can do a quick check to verify whether blocking was efficient or not. We would like the block factor to explain a lot of variation, hence if the mean square of the block factor is larger than the error mean square MS_E we conclude that blocking was efficient (compared to a completely randomized design). Here, this is the case as $794 > 150$. See Kuehl (2000) for more details and a formal definition of the relative efficiency which compares the efficiency of a randomized complete block design to a completely randomized design. If blocking was not efficient, we would still leave the block factor in the model (as the model must follow the design that we used), but we might plan not to use blocking in a *future* similar experiment because it didn't help reduce variance and only cost us degrees of freedom.

Instead of a single treatment factor, we can also have a factorial treatment structure within every block. Think for example of a design as outlined in Table 5.2.

TABLE 5.2: Block design with a factorial treatment structure with two factors A and B having two levels each (indicated in the subscript). Columns correspond to different blocks, rows to experimental units in each block.

1	2	3	...
A_1B_2	A_2B_2	A_2B_1	...
A_2B_1	A_2B_1	A_1B_2	...
A_2B_2	A_1B_2	A_1B_1	...
A_1B_1	A_1B_1	A_2B_2	...

In R, we would model this as y ~ Block + A * B. In such a situation, we can actually test the interaction between A and B even if every level combination of A and B appears only *once* in every block. Why? Because we have multiple blocks, we have multiple observations for every combination of the levels of A and B. Of course, the three-way interaction cannot be added to the model.

Interpretation of the coefficients of the corresponding models, residual analysis, etc. is done "as usual." The only difference is that we do not test the block factor for statistical significance, but for efficiency.

5.3 Nonparametric Alternatives

For a complete block design with only *one* treatment factor and no replicates, there is a rank sum based test, the so-called **Friedman rank sum test** which is implemented in function `friedman.test`. Among others, it also has a formula interface, where the block factor comes after the symbol "|", i.e., for the oat example the formula would be `yield ~ variety | block`:

```
friedman.test(yield ~ variety | block, data = oat.variety)
```

```
##
##  Friedman rank sum test
##
## data:  yield and variety and block
## Friedman chi-squared = 2.3, df = 2, p-value = 0.3
```

5.4 Outlook: Multiple Block Factors

We can also block on more than one factor. A special case is the so-called **Latin Square design** where we have *two* block factors and one treatment factor having g levels each (yes, *all* of them!). Hence, this is a *very* restrictive assumption. Consider the layout in Table 5.3 where we have a block factor with levels R_1 to R_4 ("rows"), another block factor with levels C_1 to C_4 ("columns") and a treatment factor with levels A to D (a new notation as now the letter is actually the level of the treatment factor). In a Latin Square design, each treatment (Latin letters) appears exactly *once* in each row and *once* in each column. A Latin Square design blocks on both rows and columns *simultaneously*. We also say it is a **row-column design**. If we ignore the columns of a Latin Square designs, the rows form an RCBD; if we ignore the rows, the columns form an RCBD.

TABLE 5.3: Latin Square design for $g = 4$.

	C_1	C_2	C_3	C_4
R_1	A	B	C	D
R_2	B	C	D	A
R_3	C	D	A	B
R_4	D	A	B	C

For example, if we have a factory with four machine types (treatment factor with levels A to D) such that each of four operators can only operate one machine on a single day, we could perform an experiment on four days and block on days (C_1 to C_4) and operators (R_1 to R_4) using a Latin Square design as shown in Table 5.3.

By design, a Latin Square with a treatment factor with g levels uses (only) g^2 experimental units. Hence, for small g, the degrees of freedom of the error term can be very small, see Table 5.4, leading to typically low power.

TABLE 5.4: Error degrees of freedom of Latin Square designs.

g	Error Degrees of Freedom
3	2
4	6
5	12
6	20

We can create a (random) Latin Square design in R for example with the function design.lsd of the package agricolae (de Mendiburu, 2020).

```
library(agricolae)
design.lsd(LETTERS[1:4])$sketch
```

```
##      [,1] [,2] [,3] [,4]
## [1,] "A"  "C"  "B"  "D"
## [2,] "C"  "A"  "D"  "B"
## [3,] "D"  "B"  "A"  "C"
## [4,] "B"  "D"  "C"  "A"
```

To analyze data from such a design, we use the main effects model

$$Y_{ijk} = \mu + \alpha_i + \beta_j + \gamma_k + \epsilon_{ijk}.$$

Here, the α_i's are the treatment effects and β_j and γ_k are the row- and column-specific **block effects** with the usual side constraints.

The design is balanced having the effect that our usual estimators and sums of squares are "working." In R, we would use the model formula y ~ Block1 + Block2 + Treat. We *cannot* fit a more complex model, including interaction effects, here because we do not have the corresponding replicates.

Multiple Latin Squares can also be combined or replicated. This allows for more flexibility. In that sense, Latin Square designs are useful building blocks of more complex designs, see for example Kuehl (2000).

With patients, it is common that one is not able to apply multiple treatments in parallel, but in a *sequential* manner over multiple time periods (think of comparing different painkillers: a different painkiller per day). In such a situation, we would like to block on patients ("as usual") but also on time periods, because there might be something like a learning or a fatigue effect over time. Such designs are called **crossover designs**, a standard reference is Jones and Kenward (2014). If the different periods are too close to each other, so-called **carryover effects** might be present: the medication of the previous period might still have an effect on the current period. Such effects can be incorporated into the model (if this is part of the research question). Otherwise, long enough washout periods should be used.

6

Random and Mixed Effects Models

In this chapter we use a new philosophy. Up to now, treatment effects (the α_i's) were *fixed*, unknown quantities that we tried to estimate. This means we were making a statement about a *specific*, *fixed* set of treatments (e.g., some specific fertilizers or different vaccine types). Such models are also called **fixed effects models**.

6.1 Random Effects Models

6.1.1 One-Way ANOVA

Now we use another point of view: We consider situations where treatments are **random samples** from a **large population** of treatments. This might seem quite special at first sight, but it is actually very natural in many situations. Think for example of investigating employee performance of those who were randomly selected. Another example could be machines that were randomly sampled from a large population of machines. Typically, we are interested in making a statement about some properties of the whole population and not of the observed individuals (here, employees or machines).

Going further with the machine example: Assume that every machine produces some samples whose quality we assess. We denote the observed quality of the jth sample on the ith machine with y_{ij}. We can model such data with the model

$$Y_{ij} = \mu + \alpha_i + \epsilon_{ij}, \tag{6.1}$$

where α_i is the effect of the ith machine. As the machines were drawn randomly from a large population, we assume

$$\alpha_i \text{ i.i.d. } \sim N(0, \sigma_\alpha^2).$$

We call α_i a **random effect**. Hence, (6.1) is a so-called **random effects model**. For the error term we have the usual assumption ϵ_{ij} i.i.d. $\sim N(0, \sigma^2)$. In addition, we assume that α_i and ϵ_{ij} are independent. This looks very similar to the old fixed effects model (2.4) at first sight. However, note that the α_i's in Equation (6.1) are **random variables** (reflecting the sampling mechanism of the data) and *not* fixed unknown parameters. This small change will have a large impact on the properties of the model. In addition, we have a new parameter σ_α^2 which is the variance of the random effect (here, the variance between different machines). Sometimes, such models are also called **variance components models** because of the different variances σ_α^2 and σ^2 (more complex models will have additional variances).

Let us now inspect some properties of model (6.1). The expected value of Y_{ij} is

$$E[Y_{ij}] = \mu.$$

For the variance of Y_{ij} we have

$$\text{Var}(Y_{ij}) = \sigma_\alpha^2 + \sigma^2.$$

For the correlation structure we get

$$\text{Cor}(Y_{ij}, Y_{kl}) = \begin{cases} 0 & i \neq k \\ \sigma_\alpha^2/(\sigma_\alpha^2 + \sigma^2) & i = k, j \neq l \\ 1 & i = k, j = l \end{cases}$$

Observations from different machines ($i \neq k$) are uncorrelated while observations from the *same* machine ($i = k$) are correlated. We also call the correlation within the same machine $\sigma_\alpha^2/(\sigma_\alpha^2 + \sigma^2)$ **intraclass correlation (ICC)**. It is large if $\sigma_\alpha^2 \gg \sigma^2$. A large value means that observations from the *same* "group" (here, machine) are much more similar than observations from *different* groups (machines). Three different scenarios for an example with

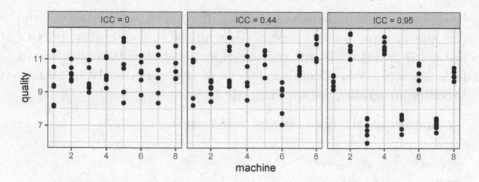

FIGURE 6.1: ICC value of three data sets of eight machines with six observations each.

six observations from each of the eight machines can be found in Figure 6.1.

This nonzero correlation within the same machine is in contrast to the fixed effects model (2.4), where all values are *independent*, because there, the α_i's are *parameters*, that is *fixed*, unknown quantities, and therefore also uncorrelated.

Parameter estimation for the variance components σ_α^2 and σ^2 is nowadays typically being done with a technique called **restricted maximum likelihood (REML)**, see for example Jiang and Nguyen (2021) for more details. We could also use classical maximum likelihood estimators here, but REML estimates are less biased. The parameter μ is estimated with maximum likelihood assuming that the variances are known. Historically, method of moments estimators were (or are) also heavily used, see for example Kuehl (2000) or Oehlert (2000).

In R, there are many packages that can fit such models. We will consider lme4 (Bates et al., 2015) and later also lmerTest (Kuznetsova et al., 2017), which basically uses lme4 for model fitting and adds some statistical tests on top.

Let us now consider Exercise 5.1 from Kuehl (2000) about an inheritance study with beef animals. Five sires, male animals, were each mated to a *separate* group of dams, female animals. The birth

weights of eight male calves (from different dams) in each of the
five sire groups were recorded. We want to find out what part of
the total variation is due to variation between different sires. In
fact, such random effects models were used very early in animal
breeding programs, see for example Henderson (1963).

We first create the data set and visualize it.

```
## Create data set ####
weight <- c(61, 100,  56, 113,  99, 103,  75,  62,   ## sire 1
            75, 102,  95, 103,  98, 115,  98,  94,   ## sire 2
            58,  60,  60,  57,  57,  59,  54, 100,   ## sire 3
            57,  56,  67,  59,  58, 121, 101, 101,   ## sire 4
            59,  46, 120, 115, 115,  93, 105,  75)   ## sire 5
sire    <- factor(rep(1:5, each = 8))
animals <- data.frame(weight, sire)
str(animals)
```

```
## 'data.frame':   40 obs. of  2 variables:
##  $ weight: num  61 100 56 113 99 103 ...
##  $ sire  : Factor w/ 5 levels "1","2","3","4",..: 1 1 1 1 ..
```

```
## Visualize data ####
stripchart(weight ~ sire, vertical = TRUE, pch = 1, xlab = "sire",
           data = animals)
```

At first sight it looks like the variation between different sires is rather small.

Now we fit the random effects model with the `lmer` function in package `lme4`. We want to have a random effect per sire. This can be specified with the notation (1 | sire) in the model formula. This means that the "granularity" of the random effect is specified after the vertical bar "|". All observations sharing the same level of `sire` will get the same random effect α_i. The "1" means that we only want to have a so-called random intercept (α_i) per sire. In the regression setup, we could also have a random slope for which we would use the notation (x | sire) for a continuous predictor x. However, this is not of relevance here.

```
library(lme4)
fit.animals <- lmer(weight ~ (1 | sire), data = animals)
```

As usual, the function `summary` gives an overview of the fitted model.

```
summary(fit.animals)
```

```
## Linear mixed model fit by REML ['lmerMod']
## ...
## Random effects:
##  Groups    Name        Variance Std.Dev.
##  sire      (Intercept) 117      10.8
##  Residual              464      21.5
## Number of obs: 40, groups:  sire, 5
##
## Fixed effects:
##             Estimate Std. Error t value
## (Intercept)    82.55       5.91      14
```

From the summary we can read off the table labelled `Random Effects` that $\hat{\sigma}_\alpha^2 = 117$ (`sire`) and $\hat{\sigma}^2 = 464$ (`Residual`). Note that the column `Std.Dev` is nothing more than the square root of the variance and *not* the standard error (accuracy) of the variance estimate.

The variance of Y_{ij} is therefore estimated as $117 + 464 = 581$. Hence, only about $117/581 = 20\%$ of the total variance of the birth weight is due to sire, this is the intraclass correlation. This confirms our visual inspection from above. It is good practice to verify whether the grouping structure was interpreted correctly. In the row Number of obs we can read off that we have 40 observations in total and a grouping factor given by sire which has five levels.

Under Fixed effects we find the estimate $\hat{\mu} = 82.55$. It is an estimate for the expected birth weight of a male calf of a randomly selected sire, randomly selected from the *whole* population of all sires.

Approximate 95% confidence intervals for the parameters can be obtained with the function confint (the argument oldNames has to bet set to FALSE to get a nicely labelled output).

```
confint(fit.animals, oldNames = FALSE)
```

```
##                        2.5 %  97.5 %
## sd_(Intercept)|sire    0.00   24.62
## sigma                 17.33   27.77
## (Intercept)           69.84   95.26
```

Hence, an approximate 95% confidence interval for the standard deviation σ_{α} is given by $[0, 24.62]$. We see that the estimate $\hat{\sigma}_{\alpha}$ is therefore quite imprecise. The reason is that we only have five sires to estimate the standard deviation. The corresponding confidence interval for the variance σ_{α}^2 would be given by $[0^2, 24.62^2]$ (we can simply apply the square function to the interval boundaries). The row labelled with (Intercept) contains the corresponding confidence interval for μ.

What would happen if we used the "ordinary" aov function here?

```
options(contrasts = c("contr.sum", "contr.poly"))
fit.animals.aov <- aov(weight ~ sire, data = animals)
confint(fit.animals.aov)
```

```
##                  2.5 % 97.5 %
## (Intercept)   75.637 89.463
## sire1        -12.751 14.901
## sire2          1.124 28.776
## sire3        -33.251 -5.599
## sire4        -18.876  8.776
```

As we have used `contr.sum`, the confidence interval for μ which can be found under (Intercept) is a confidence interval for the average of the expected birth weight of the male offsprings of these five *specific* sires. More precisely, each of the five sires has its expected birth weight $\mu_i, i = 1, \ldots, 5$ (of a male offspring). With `contr.sum` we have

$$\mu = \frac{1}{5}\sum_{i=1}^{g}\mu_i,$$

i.e., μ is the average of the expected value of these five *specific* sires. We see that this confidence interval is *shorter* than the one from the random effects model. The reason is the different interpretation. The random effects model allows to make inference about the population of all sires (where we have seen five so far), while the fixed effects model allows to make inference about these five *specific* sires. Hence, we are facing a more difficult problem with the random effects model; this is why we are less confident in our estimate resulting in wider confidence intervals compared to the fixed effects model.

We can also get "estimates" of the random effects α_i with the function `ranef`.

```
ranef(fit.animals)
```

```
## $sire
##    (Intercept)
## 1       0.7183
## 2       9.9895
## 3     -12.9797
## 4      -3.3744
## 5       5.6462
```

```
##
## with conditional variances for "sire"
```

More precisely, as the random effects are random quantities and *not* fixed parameters, what we get with `ranef` are the conditional means (given the observed data) which are the best linear unbiased predictions, also known as **BLUPs** (Robinson, 1991). Typically, these estimates are shrunken toward zero because of the normal assumption, hence we get slightly different results compared to the fixed effects model. This is also known as the so-called **shrinkage property**.

We should of course also check the model assumptions. Here, this includes in addition normality of the random effects, though this is hard to check with only five observations. We get the Tukey-Anscombe plot with the `plot` function.

```
plot(fit.animals) ## TA-plot
```

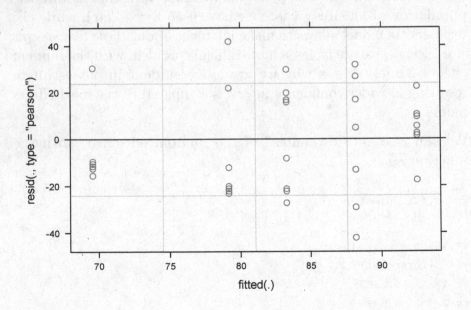

To get QQ-plots of the random effects and the residuals, we need
to extract them first and then use qqnorm as usual.

```
par(mfrow = c(1, 2))
qqnorm(ranef(fit.animals)$sire[,"(Intercept)"],
       main = "Random effects")
qqnorm(resid(fit.animals), main = "Residuals")
```

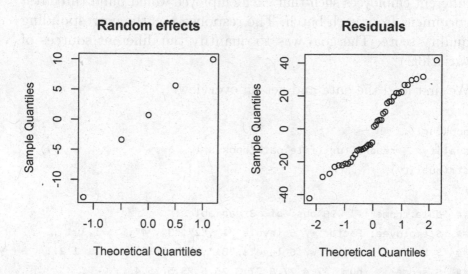

Depending on the model complexity, residual analysis for models
including random effects can be subtle, this includes the models we
will learn about in Section 6.2 too, see for example Santos Nobre
and da Motta Singer (2007), Loy and Hofmann (2015) or Loy et al.
(2017). Some implementations of these papers can be found in
package HLMdiag (Loy and Hofmann, 2014).

Sometimes you also see statistical tests of the form $H_0 : \sigma_\alpha = 0$
vs. $H_A : \sigma_\alpha > 0$. For model (6.1) we could actually get exact p-
values, while for complex models later this is not the case anymore.
In these situations, approximate or simulation based methods have
to be used. Some can be found in package RLRsim (Scheipl et al.,
2008).

6.1.2   More Than One Factor

So far this was a one-way ANOVA model with a random effect. We can extend this to the two-way ANOVA situation and beyond. For this reason, we consider the following example:

A large company randomly selected five employees and six batches of source material from the production process. The material from each batch was divided into 15 pieces which were randomized to the different employees such that each employee would build three test specimens from each batch. The response was the corresponding quality score. The goal was to quantify the different sources of variation.

We first load the data and get an overview.

```
book.url <- "https://stat.ethz.ch/~meier/teaching/book-anova"
quality <- readRDS(url(file.path(book.url, "data/quality.rds")))
str(quality)
```

```
## 'data.frame':     90 obs. of  3 variables:
## $ employee: Factor w/ 5 levels "1","2","3","4",..: 1 1 1 ..
## $ batch    : Factor w/ 6 levels "B1","B2","B3",..: 1 1 1 2..
## $ score    : num  27.4 27.8 27.3 25.5 25.5 26.4 ...
```

```
xtabs(~ batch + employee, data = quality)
```

```
##       employee
## batch 1 2 3 4 5
##    B1 3 3 3 3 3
##    B2 3 3 3 3 3
##    B3 3 3 3 3 3
##    B4 3 3 3 3 3
##    B5 3 3 3 3 3
##    B6 3 3 3 3 3
```

We can for example visualize this data set with an interaction plot.

```
with(quality, interaction.plot(x.factor = batch,
                               trace.factor = employee,
                               response = score))
```

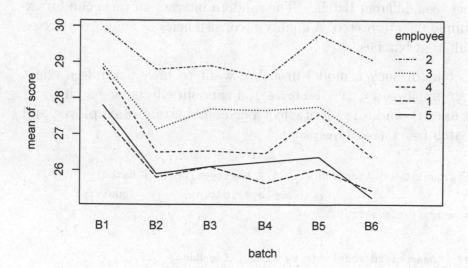

There seems to be variation between different employees and between the different batches. The interaction does not seem to be very pronounced. Let us set up a model for this data. We denote the observed quality score of the kth sample of employee i using material from batch j by y_{ijk}. A natural model is then

$$Y_{ijk} = \mu + \alpha_i + \beta_j + (\alpha\beta)_{ij} + \epsilon_{ijk}, \qquad (6.2)$$

where α_i is the random (main) effect of employee i, β_j is the random (main) effect of batch j, $(\alpha\beta)_{ij}$ is the random interaction term between employee i and batch j and ϵ_{ijk} i.i.d. $\sim N(0, \sigma^2)$ is the error term. For the random effects we have the usual assumptions

$$\alpha_i \text{ i.i.d. } \sim N(0, \sigma_\alpha^2),$$
$$\beta_j \text{ i.i.d. } \sim N(0, \sigma_\beta^2),$$
$$(\alpha\beta)_{ij} \text{ i.i.d. } \sim N(0, \sigma_{\alpha\beta}^2).$$

In addition, we assume that α_i, β_j, $(\alpha\beta)_{ij}$ and ϵ_{ijk} are independent. As each random term in the model has its own variance component, we now have the variances σ_α^2, σ_β^2, $\sigma_{\alpha\beta}^2$ and σ^2.

What is the interpretation of the different terms? The random (main) effect of employee is the variability between different employees, and the random (main) effect of batch is the variability between different batches. The random interaction term can for example be interpreted as quality inconsistencies of employees across different batches.

Let us fit such a model in R. We want to have a random effect per employee (= (1 | employee)), a random effect per batch (= (1 | batch)), and a random effect per combination of employee and batch (= (1 | employee:batch)).

```
fit.quality <- lmer(score ~ (1 | employee) + (1 | batch) +
                    (1 | employee:batch), data = quality)
summary(fit.quality)
```

```
## Linear mixed model fit by REML ['lmerMod']
## ...
##
## Random effects:
##  Groups          Name        Variance Std.Dev.
##  employee:batch (Intercept) 0.0235   0.153
##  batch          (Intercept) 0.5176   0.719
##  employee       (Intercept) 1.5447   1.243
##  Residual                   0.2266   0.476
## Number of obs: 90, groups:
## employee:batch, 30; batch, 6; employee, 5
##
## Fixed effects:
##             Estimate Std. Error t value
## (Intercept)   27.248      0.631    43.2
```

From the output we get $\hat{\sigma}_\alpha^2 = 1.54$ (employee), $\hat{\sigma}_\beta^2 = 0.52$ (batch), $\hat{\sigma}_{\alpha\beta}^2 = 0.02$ (interaction of employee and batch) and $\hat{\sigma}^2 = 0.23$ (error term).

Hence, total variance is $1.54 + 0.52 + 0.02 + 0.23 = 2.31$. We see that the largest contribution to the variance is variability between different employees which contributes to about $1.54/2.31 \approx 67\%$ of the total variance. These are all point estimates. Confidence intervals on the scale of the standard deviations can be obtained by calling confint.

```
confint(fit.quality, oldNames = FALSE)
```

```
##                                    2.5 %    97.5 %
## sd_(Intercept)|employee:batch    0.0000    0.3528
## sd_(Intercept)|batch             0.4100    1.4754
## sd_(Intercept)|employee          0.6694    2.4994
## sigma                            0.4020    0.5741
## (Intercept)                     25.9190   28.5766
```

Uncertainty is quite large. In addition, there is no statistical evidence that the random interaction term is really needed, as the corresponding confidence interval contains zero. To get a more narrow confidence interval for the standard deviation between different employees, we would need to sample more employees. Basically, each employee contributes one observation for estimating the standard deviation (or variance) between employees. The same reasoning applies to batch.

6.1.3 Nesting

To introduce a new concept, we consider the Pastes data set in package lme4. The strength of a chemical paste product was measured for a total of 30 samples coming from 10 randomly selected delivery batches where each contained three randomly selected casks. Each sample was measured twice, resulting in a total of 60 observations.

We want to check what part of the total variability of strength is due to variability between batches, between casks and due to measurement error.

```
data("Pastes", package = "lme4")
str(Pastes)
```

```
## 'data.frame':     60 obs. of  4 variables:
##  $ strength: num  62.8 62.6 60.1 62.3 62.7 63.1 ...
##  $ batch   : Factor w/ 10 levels "A","B","C","D",..: 1 1 1..
##  $ cask    : Factor w/ 3 levels "a","b","c": 1 1 2 2 3 3 ...
##  $ sample  : Factor w/ 30 levels "A:a","A:b","A:c",..: 1 1..
```

Note that the levels of batch and cask are given by upper- and lowercase letters, respectively. If we carefully think about the data structure, we have just discovered a new way of combining factors. Cask a in batch A has nothing to do with cask a in batch B and so on. The level a of cask has a different meaning for every level of batch. Hence, the two factors cask and batch are *not* crossed. We say cask is **nested** in batch. The data set also contains an additional (redundant) factor sample which is a unique identifier for each sample, given by the combination of batch and cask.

We use package ggplot2 to visualize the data set (R code for interested readers only). The different panels are the different batches (A to J).

```
library(ggplot2)
ggplot(Pastes, aes(y = cask, x = strength)) + geom_point() +
  facet_grid(batch ~ .)
```

The batch effect does not seem to be very pronounced, for example, there is no clear tendency that some batches only contain large values, while others only contain small values. Casks within the same batch can be substantially different, but the two measurements from the same cask are typically very similar. Let us now set up an appropriate random effects model for this data set.

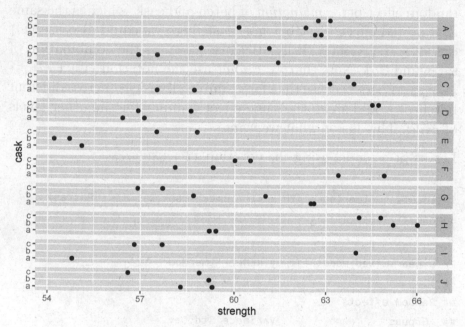

Let y_{ijk} be the observed strength of the kth sample of cask j in batch i. We can then use the model

$$Y_{ijk} = \mu + \alpha_i + \beta_{j(i)} + \epsilon_{ijk}, \qquad (6.3)$$

where α_i is the random effect of batch and $\beta_{j(i)}$ is the random effect of cask *within* batch. Note the special notation $\beta_{j(i)}$ which emphasizes that cask is nested in batch. This means that each batch gets its *own* coefficient for the cask effect (from a technical point of view this is like an interaction effect without the corresponding main effect). We make the usual assumptions for the random effects

$$\alpha_i \text{ i.i.d.} \sim N(0, \sigma_\alpha^2), \quad \beta_{j(i)} \text{ i.i.d.} \sim N(0, \sigma_\beta^2),$$

and similarly for the error term $\epsilon_{ijk} \sim N(0, \sigma^2)$. As before, we assume independence between all random terms.

We have to tell lmer about the nesting structure. There are multiple ways to do so. We can use the notation (1 | batch/cask) which means that we want to have a random effect per batch and per cask *within* batch. We could also use (1 | batch) + (1 | cask:batch) which means that we want to have a random effect per batch and a random effect per *combination* of batch and cask, which is the same as a nested effect. Last but not least, here we could also use (1 | batch) + (1 | sample). Why? Because sample is the combination of batch and cask. Note that we *cannot* use the notation (1 | batch) + (1 | cask) here because then for example all casks a (across different batches) would share the same effect (similar for the other levels of cask). This is not what we want.

Let us fit a model using the notation (1 | batch/cask).

```
fit.paste <- lmer(strength ~ (1 | batch/cask), data = Pastes)
summary(fit.paste)
```

```
## Linear mixed model fit by REML ['lmerMod']
## ...
## Random effects:
##  Groups      Name        Variance  Std.Dev.
##  cask:batch  (Intercept) 8.434     2.904
##  batch       (Intercept) 1.657     1.287
##  Residual                0.678     0.823
## Number of obs: 60, groups:  cask:batch, 30; batch, 10
##
## Fixed effects:
##             Estimate Std. Error t value
## (Intercept)  60.053     0.677      88.7
```

We have $\hat{\sigma}_\alpha^2 = 1.66$ (batch), $\hat{\sigma}_\beta^2 = 8.43$ (cask) and $\hat{\sigma}^2 = 0.68$ (measurement error). This confirms that most variation is in fact due to cask within batch. Confidence intervals could be obtained as usual with the function confint (not shown).

A nested effect has also some associated degrees of freedom. In the previous example, cask has 3 levels in each of the 10 batches. The reasoning is now as follows: As cask needs 2 degrees of freedom in each level of the 10 batches, the degrees of freedom of cask are $10 \cdot 2 = 20$.

6.2 Mixed Effects Models

In practice, we often encounter models which contain both random and fixed effects. We call them **mixed models** or **mixed effects models**.

6.2.1 Example: Machines Data

We start with the data set `Machines` in package `nlme` (Pinheiro et al., 2021). As stated in the help file: "Data on an experiment to compare three brands of machines used in an industrial process [...]. Six workers were chosen randomly among the employees of a factory to operate each machine three times. The response is an overall productivity score taking into account the number and quality of components produced."

```
data("Machines", package = "nlme")
## technical details for shorter output:
class(Machines) <- "data.frame"
Machines[, "Worker"] <- factor(Machines[, "Worker"], levels = 1:6,
                              ordered = FALSE)
str(Machines, give.attr = FALSE) ## give.attr to shorten output
```

```
## 'data.frame':    54 obs. of  3 variables:
##  $ Worker : Factor w/ 6 levels "1","2","3","4",..: 1 1 1 2..
##  $ Machine: Factor w/ 3 levels "A","B","C": 1 1 1 1 1 1 ...
##  $ score  : num  52 52.8 53.1 51.8 52.8 53.1 ...
```

Let us first visualize the data. In addition to plotting all individual data, we also calculate the mean for each combination of worker and machine and connect these values with lines for each worker (R code for interested readers only).

```
ggplot(Machines, aes(x = Machine, y = score, group = Worker,
                    shape = Worker, linetype = Worker)) +
```

```
geom_point() + stat_summary(fun = mean, geom = "line") +
theme_bw()
```

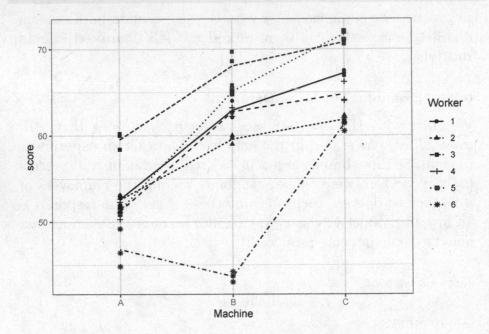

```
## A classical interaction plot would be (not shown)
with(Machines, interaction.plot(x.factor = Machine,
                                trace.factor = Worker,
                                response = score))
```

We observe that on average, productivity is largest on machine C, followed by B and A. Most workers show a similar profile, with the exception of worker 6 who performs badly on machine B.

Let us now try to model this data. The goal is to make inference about the specific machines at hand, this is why we treat machine as a fixed effect. We assume that there is a population machine effect (think of an average profile across *all* potential workers), but each worker is allowed to have its own random deviation. With y_{ijk} we denote the observed kth productivity score of worker j on machine i. We use the model

$$Y_{ijk} = \mu + \alpha_i + \beta_j + (\alpha\beta)_{ij} + \epsilon_{ijk}, \tag{6.4}$$

where α_i is the fixed effect of machine i (with a usual side constraint), β_j is the random effect of worker j and $(\alpha\beta)_{ij}$ is the corresponding random interaction effect. An interaction effect between a random (here, worker) and a fixed effect (here, machine) is treated as a random effect.

What is the interpretation of this model? The average (over the whole population) productivity profile with respect to the three different machines is given by the fixed effect α_i. The random deviation of an individual worker consists of a general shift β_j (worker specific *general* productivity level) and a worker specific preference $(\alpha\beta)_{ij}$ with respect to the three machines. We assume that all random effects are normally distributed, this means

$$\beta_j \text{ i.i.d.} \sim N(0, \sigma_\beta^2), \quad (\alpha\beta)_{ij} \text{ i.i.d.} \sim N(0, \sigma_{\alpha\beta}^2).$$

For the error term we assume as always ϵ_{ijk} i.i.d. $\sim N(0, \sigma^2)$. In addition, all random terms are assumed to be independent.

We visualize model (6.4) step-by-step in Figure 6.2. The solid line is the population average of the machine effect ($= \mu + \alpha_i$), the dashed line is the population average including the worker specific general productivity level ($= \mu + \alpha_i + \beta_j$), i.e., a *parallel* shift of the solid line. The dotted line in addition includes the worker specific machine preference ($= \mu + \alpha_i + \beta_j + (\alpha\beta)_{ij}$). Individual observations of worker j now fluctuate around these values.

We could use the lme4 package to fit such a model. Besides the fixed effect of machine type we want to have the following random effects:

- a random effect β_j per worker: (1 | Worker)
- a random effect $(\alpha\beta)_{ij}$ per combination of worker and machine: (1 | Worker:Machine)

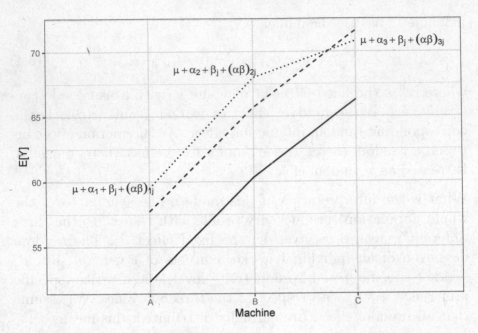

FIGURE 6.2: Illustration of model (6.4) for the machines data set.

Hence, the `lmer` call would look as follows.

```
fit.machines <- lmer(score ~ Machine + (1 | Worker) +
                    (1 | Worker:Machine), data = Machines)
```

As `lme4` does *not* calculate p-values for the fixed effects, we use the package `lmerTest` instead. Technically speaking, `lmerTest` uses `lme4` to fit the model and then adds some statistical tests, i.e., p-values for the fixed effects, to the output. There are still many open issues regarding statistical inference in mixed models, see for example the "GLMM FAQ"[1] (the so-called generalized linear mixed models frequently asked questions). However, for "nice" designs (as the current one), we get proper statistical inference using the `lmerTest` package.

---

[1] http://bbolker.github.io/mixedmodels-misc/glmmFAQ.html

```
options(contrasts = c("contr.treatment", "contr.poly"))
library(lmerTest)
fit.machines <- lmer(score ~ Machine + (1 | Worker) +
                     (1 | Worker:Machine), data = Machines)
```

We get the ANOVA table for the fixed effects with the function anova.

```
anova(fit.machines)
```

```
## Type III Analysis of Variance Table with Satterthwaite's method
##           Sum Sq Mean Sq NumDF DenDF F value  Pr(>F)
## Machine    38.1     19     2    10   20.6 0.00029
```

The fixed effect of machine is significant. If we closely inspect the output, we see that an F-distribution with 2 and 10 degrees of freedom is being used. Where does the value 10 come from? It seems to be a rather small number given the sample size of 54 observations. Let us remind ourselves what the fixed effect actually means. It is the *average* machine effect, where the average is taken over the whole population of workers (where we have seen only six). We know that every worker has its own random deviation of this effect. Hence, the relevant question is whether the worker profiles just fluctuate around a constant machine effect (all $\alpha_i = 0$) or whether the machine effect is substantially larger than this worker specific machine variation. Technically speaking, the worker specific machine variation is nothing more than the *interaction* between worker and machine. Hence, this boils down to comparing the variation between different machines (having 2 degrees of freedom) to the variation due to the interaction between machines and workers (having $2 \cdot 5 = 10$ degrees of freedom). The lmerTest package automatically detects this because of the structure of the random effects. This way of thinking also allows another insight: If we want the estimate of the population average of the machine effect (the fixed effect of machine) to have a desired accuracy, the relevant quantity to increase is the number of workers.

If we call summary on the fitted object, we get the individual $\hat{\alpha}_i$'s under Fixed effects.

```
summary(fit.machines)
```

```
## Linear mixed model fit by REML. t-tests use
##    Satterthwaite's method [lmerModLmerTest]
## ...
##
## Random effects:
##  Groups          Name        Variance Std.Dev.
##  Worker:Machine (Intercept) 13.909   3.730
##  Worker         (Intercept) 22.858   4.781
##  Residual                    0.925   0.962
## Number of obs: 54, groups:  Worker:Machine, 18; Worker, 6
##
## Fixed effects:
##             Estimate Std. Error    df t value Pr(>|t|)
## (Intercept)    52.36       2.49  8.52   21.06  1.2e-08
## MachineB        7.97       2.18 10.00    3.66   0.0044
## MachineC       13.92       2.18 10.00    6.39  7.9e-05
## ...
```

Let us first check whether the structure was interpreted correctly (Number of obs): There are 54 observations, coming from 6 different workers and 18 ($= 3 \cdot 6$) different combinations of workers and machines.

As usual, we have to be careful with the interpretation of the fixed effects because it depends on the side constraint that is being used. For example, here we can read off the output that the productivity score on machine B is on average 7.97 units larger than on machine A. We can also get the parameter estimates of the fixed effects only by calling fixef(fit.machines) (not shown).

Estimates of the different variance components can be found under Random effects. Often, we are not very much interested in the actual values. We rather use the random effects as a "tool" to model correlated data and to tell the fitting function that we are interested in the population average and *not* the worker specific machine effects. In that sense, including random effects in a model changes the interpretation of the fixed effects. Hence, it really depends on the research question whether we treat a factor as fixed or random. There is no right or wrong here.

Approximate 95% confidence intervals can be obtained as usual with the function `confint`.

```
confint(fit.machines, oldNames = FALSE)
```

```
##                                 2.5 % 97.5 %
## sd_(Intercept)|Worker:Machine   2.353  5.432
## sd_(Intercept)|Worker           1.951  9.411
## sigma                           0.776  1.235
## (Intercept)                    47.395 57.316
## MachineB                        3.738 12.195
## MachineC                        9.688 18.145
```

For example, a 95% confidence interval for the expected value of the difference between machine A and B is given by $[3.7, 12.2]$.

We can do the residual analysis as outlined in Section 6.1.1.

```
## Tukey-Anscombe plot
plot(fit.machines)
```

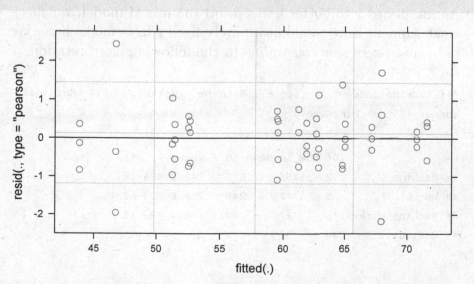

```
## QQ-plots
par(mfrow = c(1, 3))
qqnorm(ranef(fit.machines)$Worker[, 1],
       main = "Random effects of worker")
qqnorm(ranef(fit.machines)$'Worker:Machine'[, 1],
       main = "Random interaction")
qqnorm(resid(fit.machines), main = "Residuals")
```

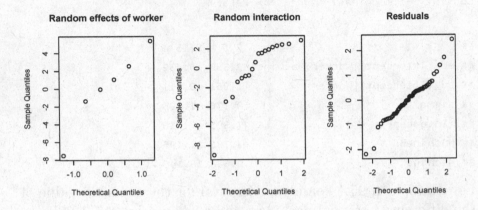

The Tukey-Anscombe plot looks good. The QQ-plots could look better; however, we do not have a lot of observations such that the deviations are still OK. Or in other words, it is difficult to detect clear violations of the normality assumption.

Again, in order to better understand the mixed model, we check what happens if we would fit a purely fixed effects model here. We use sum-to-zero side constraints in the following interpretation.

```
fit.machines.aov <- aov(score ~ Machine * Worker, data = Machines)
summary(fit.machines.aov)
```

##	Df	Sum Sq	Mean Sq	F value	Pr(>F)
## Machine	2	1755	878	949.2	<2e-16
## Worker	5	1242	248	268.6	<2e-16
## Machine:Worker	10	427	43	46.1	<2e-16
## Residuals	36	33	1		

The machine effect is much more significant. This is because in the fixed effects model, the main effect of machine makes a statement about the average machine effect of these 6 *specific* workers and *not* about the population average (in the same spirit as in the sire example in Section 6.1.1).

Remark: The model in Equation (6.4) with $(\alpha\beta)_{ij}$ i.i.d. $\sim$ $N(0, \sigma^2_{\alpha\beta})$ is also known as the so-called **unrestricted model**. There is also a **restricted model** where we would assume that a random interaction effect sums up to zero if we sum over the *fixed* indices. For our example this would mean that if we sum up the interaction effect for each worker across the different machines, we would get zero. Here, the restricted model means that the random interaction term does not contain any information about the general productivity level of a worker, just about machine preference. Unfortunately, the restricted model is currently not implemented in lmer. Luckily, the inference with respect to the *fixed* effects is the same for both models, only the estimates of the variance components differ. For more details, see for example Burdick and Graybill (1992).

6.2.2   Example: Chocolate Data

We continue with a more complex design which is a relabelled version of Example 12.2 of Oehlert (2000) (with some simulated data). A group of 10 raters with rural background and 10 raters with urban background rated 4 different chocolate types. Every rater tasted and rated two samples from the same chocolate type in random order. Hence, we have a total of $20 \cdot 4 \cdot 2 = 160$ observations.

```
book.url <- "http://stat.ethz.ch/~meier/teaching/book-anova"
chocolate <- read.table(file.path(book.url, "data/chocolate.dat"),
                        header = TRUE)
chocolate[,"rater"]      <- factor(chocolate[,"rater"])
chocolate[,"background"] <- factor(chocolate[,"background"])
str(chocolate)
```

```
## 'data.frame':   160 obs. of  4 variables:
```

```
##  $ choc      : chr  "A" "A" ...
##  $ rater     : Factor w/ 10 levels "1","2","3","4",..: 1 1..
##  $ background: Factor w/ 2 levels "rural","urban": 1 1 1 1..
##  $ y         : int  61 64 46 45 63 66 ...
```

Chocolate type is available in factor choc (with levels A, B, C and D), background of rater (with levels rural and urban) in background, the score in y and a rater ID can be found in rater. Note that the factor rater has only 10 levels. We have to be careful here and should not forget that rater is actually *nested* in background; so, the raters labelled 1 in the urban and the rural group have *nothing* in common.

We can easily visualize this data with an interaction plot. We use the package ggplot2 to get a more appealing plot compared to the function interaction.plot (R code for interested readers only).

```
ggplot(chocolate, aes(x = choc, y = y,
                      group = interaction(background, rater),
                      linetype = background)) +
  stat_summary(fun = mean, geom = "line") + theme_bw()
```

Questions that could arise are:

- Is there a background effect (urban vs. rural)?
- Is there a chocolate type effect?
- Does the effect of chocolate type depend on background (interaction)?
- How large is the variability between different raters regarding general chocolate liking level?
- How large is the variability between different raters regarding chocolate type preference?

Here, background and chocolate type are fixed effects, as we want to make a statement about these *specific* levels. However, we treat rater as random, as we want the fixed effects to be population average effects and we are interested in the variation between different raters.

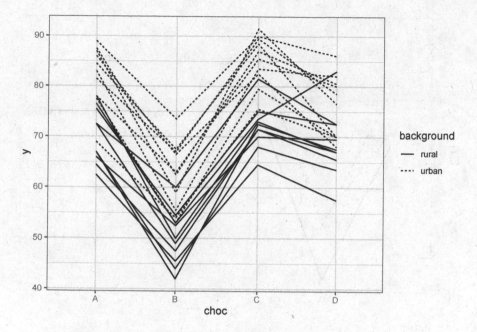

FIGURE 6.3: Interaction plot of the chocolate data set.

We can use the model

$$Y_{ijkl} = \mu + \alpha_i + \beta_j + (\alpha\beta)_{ij} + \delta_{k(i)} + (\beta\delta)_{jk(i)} + \epsilon_{ijkl}, \quad (6.5)$$

where

- α_i is the (fixed) effect of background, $i = 1, 2$,
- β_j is the (fixed) effect of chocolate type, $j = 1, \ldots, 4$,
- $(\alpha\beta)_{ij}$ is the corresponding interaction (fixed effect),
- $\delta_{k(i)}$ is the random effect of rater: "general chocolate liking level of rater $k(i)$ (nested in background!)", $k = 1, \ldots, 10$,
- $(\beta\delta)_{jk(i)}$ is the (random) interaction of rater and chocolate type: "specific chocolate type preference of rater $k(i)$".

We use the usual assumptions for all random terms. In this setup, for both backgrounds there is (over the whole population) an average preference profile with respect to the 4 different chocolate types (given by $\mu + \alpha_i + \beta_j + (\alpha\beta)_{ij}$). Every rater can have its individual random deviation from this profile. This deviation consists of a general shift $\delta_{k(i)}$ (general chocolate liking of a rater) and a rater

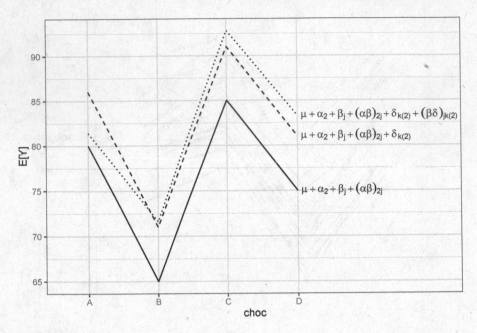

FIGURE 6.4: Illustration of model (6.5) for the chocolate data set for urban raters.

specific preference $(\beta\delta)_{jk(i)}$ with respect to the 4 chocolate types. This is visualized in Figure 6.4 (for urban raters only, $i = 2$). The situation is very similar to the previous example about machines in Section 6.2.1. The solid line is the population average ($= \mu + \alpha_2 + \beta_j + (\alpha\beta)_{2j}$). The dashed line is a parallel shift including the rater specific general chocolate liking level ($= \mu + \alpha_2 + \beta_j + (\alpha\beta)_{2j} + \delta_{k(2)}$). The dotted line in addition includes the rater specific chocolate type preference ($\mu + \alpha_2 + \beta_j + (\alpha\beta)_{2j} + \delta_{k(2)} + (\beta\delta)_{jk(2)}$). Therefore, the fixed effects have to be interpreted as population averages.

Let us now try to fit this model in R. We want to have

- main effects and the interaction for the fixed effects of background and chocolate type: `background * choc`
- a random effect per rater (nested in background): `(1 | rater:background)`
- a random effect per chocolate type and rater (nested in background): `(1 | rater:background:choc)`.

We always write `rater:background` to get a unique rater ID. An alternative would be to define another factor in the data set which enumerates the raters from 1 to 20 (instead from 1 to 10).

Hence, the `lmer` call (using package `lmerTest`) looks as follows.

```
fit.choc <- lmer(y ~ background * choc + (1 | rater:background) +
                 (1 | rater:background:choc), data = chocolate)
```

As before, we get an ANOVA table with p-values for the fixed effects with the function `anova`.

```
anova(fit.choc)
```

```
## Type III Analysis of Variance Table with Satterthwaite's method
##                  Sum Sq Mean Sq NumDF DenDF F value  Pr(>F)
## background          263     263     1    18   27.61 5.4e-05
## choc               4219    1406     3    54  147.74 < 2e-16
## background:choc      64      21     3    54    2.24   0.094
```

We see that the interaction is not significant but both main effects are. This is what we already observed in the interaction plot in Figure 6.3. There, the profiles were quite parallel, but raters with an urban background rated higher on average than those with a rural background. In addition, there was a clear difference between different chocolate types.

Can we get an intuitive idea about the denominator degrees of freedom that are being used in these tests?

- The main effect of background can be thought of as a two-sample t-test with two groups having 10 observations each. Think of taking one average value per rater. Hence, we get $2 \cdot 10 - 2 = 18$ degrees of freedom.
- As we allow for a rater specific chocolate type preference, we have to check whether the effect of chocolate is substantially larger than this rater specific variation. The rater specific variation can be thought of as the interaction between rater and chocolate type. As rater is nested in background, rater has 9 degrees of

freedom in each background group. Hence, the interaction has
$((9+9)\cdot 3 = 54)$ degrees of freedom.

- The same argument as above holds true for the interaction
between background and chocolate type.

Hence, if we want to get a more precise view about these population
average effects, the relevant quantity to increase is the number of
raters.

Approximate confidence intervals for the individual coefficients
of the fixed effects and the variance components could again be
obtained by calling `confint(fit.choc, oldNames = FALSE)` (output not
shown).

Most often, the focus is not on the actual values of the variance
components. We use these random effects to change the meaning
of the fixed effects. By incorporating the random interaction effect
between rater and chocolate type, the meaning of the fixed effects
changes. We have seen 20 different "profiles" (one for each rater).
The fixed effects must now be understood as the corresponding population average effects. This also holds true for the corresponding
confidence intervals. They all make a statement about a population
average.

Remark: Models like this and the previous ones were of course
fitted before specialized mixed model software was available. For
such a nicely balanced design as we have here, we could in fact use
the good old `aov` function. In order to have more readable R code,
we first create a new variable `unique.rater` which will have 20 levels,
one for each rater (we could have done this for `lmer` too, see the
comment above).

```
chocolate[,"unique.rater"] <- with(chocolate,
                                   interaction(background, rater))
str(chocolate)
```

```
## 'data.frame':    160 obs. of  5 variables:
##  $ choc        : chr  "A" "A" ...
##  $ rater       : Factor w/ 10 levels "1","2","3","4",..: 1..
```

```
##  $ background  : Factor w/ 2 levels "rural","urban": 1 1 1..
##  $ y           : int  61 64 46 45 63 66 ...
##  $ unique.rater: Factor w/ 20 levels "rural.1","urban.1",...
```

Now we have to tell the aov function that it should use different so-called error strata. We can do this by adding Error() to the model formula. Here, we use Error(unique.rater/choc). This is equivalent to adding a random effect for each rater (1 | unique.rater) and a rater-specific chocolate effect (1 | unique.rater:choc).

```
fit.choc.aov <- aov(y ~ background * choc +
                    Error(unique.rater/choc), data = chocolate)
summary(fit.choc.aov)
```

```
##
## Error: unique.rater
##             Df Sum Sq Mean Sq F value  Pr(>F)
## background   1   5119    5119    27.6 5.4e-05
## Residuals   18   3337     185
##
## Error: unique.rater:choc
##                 Df Sum Sq Mean Sq F value Pr(>F)
## choc             3  12572    4191  147.74 <2e-16
## background:choc  3    190      63    2.24  0.094
## Residuals       54   1532      28
## ...
```

We simply put the random effects into the Error() term and get the same results as with lmer. However, the approach using lmer is much more flexible, especially for unbalanced data.

6.2.3 Outlook

Both the machines and the chocolate data were examples of so-called **repeated measurement data**. Repeated measurements occur if we have multiple measurements of the response variable from each experimental unit, e.g., over time, space, experimental conditions, etc. In the last two examples, we had observations across different experimental conditions (different machines or chocolate types). For such data, we distinguish between so-called between-subjects and within-subjects factors.

A **between-subjects factor** is a factor that splits the subjects into different groups. For the machines data, such a thing does not exist. However, for the chocolate data, background is a between-subjects factor because it splits the 20 raters into two groups: 10 with rural and 10 with urban background. On the other hand, a **within-subjects factor** splits the observations from an individual subject into different groups. For the machines data, this is the machine brand (factor `Machine` with levels `A`, `B` and `C`). For the chocolate data, this is the factor `choc` with levels `A`, `B`, `C` and `D`. Quite often, the within-subjects factor is time, e.g., when investigating growth curves.

Why are such designs popular? They are of course needed if we are interested in individual, subject-specific, patterns. In addition, these designs are efficient because for the within-subjects factors we block on subjects. Subjects can even serve as their own control!

The `lmer` model formula of the corresponding models all follow a similar pattern. As a general rule, we always include a random effect for each subject (`1 | subject`). This tells `lmer` that the values from an individual subject belong together and are therefore correlated. If treatment is a within-subjects factor and if we have multiple observations from the same treatment for each subject, we will typically introduce another random effect (`1 | subject:treatment`) as we did for the chocolate data. This means that the effect of the treatment is slightly different for each subject. Hence, the corresponding fixed effect of treatment has to be interpreted as the population average effect. If we do not have replicates of the same treatment within the subjects, we would just use the random effect per subject, that is (`1 | subject`).

If treatment is a between-subjects factor (meaning we randomize treatments to subjects) and if we track subjects across different time-points (with one observation for each time-point and subject) we could use a model of the form `y ~ treatment * time + (1 | subject)` where `time` is treated as a factor; we are going to see this model again in Chapter 7. Here, if we need the interaction between `treatment` and `time` it means that the time-development is treatment-specific. Or in other words, each treatment would have its own profile with respect to time. An alternative approach would be to aggregate the values of each subject (a short time-series) into a single meaningful number determined by the research question, e.g., slope, area under the curve, time to peak, etc. The analysis is then much easier as we only have one observation for each subject and we are back to a completely randomized design, see Chapter 2. This means we would not need any random effects anymore and `aov(aggregated.response ~ treatment)` would be enough for the analysis. Such an approach is also called a **summary statistic** or **summary measure analysis** (very easy and very powerful!). For more details about the analysis of repeated measurements data, see for example Fitzmaurice et al. (2011).

In general, one has to be careful whether the corresponding statistical tests are performed correctly. For example, if we have a between-subjects factor like background for the chocolate data, the relevant sample size is the number of subjects in each group, and not the number of measurements made on all subjects. This is why the test for background only used 18 denominator degrees of freedom. Depending on the software that is being used, these tests are not always performed correctly. It is therefore advisable to perform some sanity checks.

We have only touched the surface of what can be done with the package `lme4` (and `lmerTest`). For example, we assumed independence between the different random effects. When using a term like `(1 | treatment:subject)`, each subject gets independent random treatment effects all having the same variance. One can also use `(treatment | subject)` in the model formula. This allows not only for correlation between the subject specific treatment effects, but also for different variance components, one component for each level of `treatment`. A technical overview also covering implementation issues can be found in Bates et al. (2015). With repeated measurements data, it is quite natural that we are faced with serial (temporal) correlation. Implementations of a lot of correlation structures can be found in package nlme (Pinheiro et al., 2021). The corresponding book (Pinheiro and Bates, 2009) contains many examples. Some implementations can also be found in package glmmTMB (Brooks et al., 2017).

7

Split-Plot Designs

In this chapter we are going to learn something about experimental designs that contain experimental units of different sizes, with different randomizations. These so-called split-plot designs are maybe the most misunderstood designs in practice; therefore, they are often analyzed in a wrong way. Unfortunately, split-plot designs are very common, although they are not always conducted on purpose or consciously. The good message is that once you know how to detect these designs, the analysis is straightforward, we just have to add the proper random effects to the model.

7.1 Introduction

We start with a fictional example. Farmer John has eight different plots of land. He randomizes and applies two fertilization schemes, control and new, in a balanced way to the eight plots. In addition, each plot is divided into four subplots. In each plot, four different strawberry varieties are randomized to the subplots. John is interested in the effect of fertilization scheme and strawberry variety on fruit mass. Per subplot, he records the fruit mass after a certain amount of time. This means we have a total of $8 \cdot 4 = 32$ observations.

The design is outlined in Table 7.1. There, the eight plots appear in the different columns. The fertilization scheme is denoted by control (ctrl) and new and as shaded text. A subplot is denoted by an individual cell in a column. Strawberry varieties are labelled using A to D. In the actual layout, the eight plots were not located side-by-side.

TABLE 7.1: Layout of the example of farmer John.

1	2	3	4	5	6	7	8
ctrl	ctrl	new	ctrl	new	ctrl	new	new
D	A	B	C	B	A	D	A
A	D	C	D	A	D	C	B
C	B	A	A	C	C	A	D
B	C	D	B	D	B	B	C

We first read the data and check the structure.

```
book.url <- "http://stat.ethz.ch/~meier/teaching/book-anova"
john <- readRDS(url(file.path(book.url, "data/john.rds")))
str(john)
```

```
## 'data.frame':    32 obs. of  4 variables:
##  $ plot      : Factor w/ 8 levels "1","2","3","4",..: 1 1 ..
##  $ fertilizer: Factor w/ 2 levels "control","new": 1 1 1 1..
##  $ variety   : Factor w/ 4 levels "A","B","C","D": 1 2 3 4..
##  $ mass      : num  8.9 9.5 11.7 15 10.8 11 ...
```

Information about the two treatment variables can be found in factor `fertilizer`, with levels `control` and `new`, and factor `variety`, with levels A, B, C and D. The response is given by the numerical variable `mass`. Here, we also have information about the plot of land in factor `plot`, with levels 1 to 8, which will be useful to better understand the design and for the model specification later on.

If we consider the two treatment factors `fertilizer` and `variety`, the design looks like a "classical" factorial design at first sight.

```
xtabs(~ fertilizer + variety, data = john)
```

```
##            variety
## fertilizer A B C D
##    control 4 4 4 4
##    new     4 4 4 4
```

When considering `plot` and `fertilizer`, we see that for example plot 1 only contains level `control` of `fertilizer`, while plot 3 only contains level `new`. We get the same pattern as in Table 7.1.

```
xtabs(~ fertilizer + plot, data = john)
```

```
##           plot
## fertilizer 1 2 3 4 5 6 7 8
##    control 4 4 0 4 0 4 0 0
##    new     0 0 4 0 4 0 4 4
```

Note that we can typically *not* recover the randomization procedure from a data-frame alone. We really need the whole "story".

We can visualize the data with an interaction plot which shows that mass is larger on average with the new fertilization scheme. In addition, there seems to be a variety effect. The interaction is not very pronounced (the variety effect seems to be consistent across the two fertilization schemes).

```
with(john, interaction.plot(x.factor = variety,
                            trace.factor = fertilizer,
                            response = mass))
```

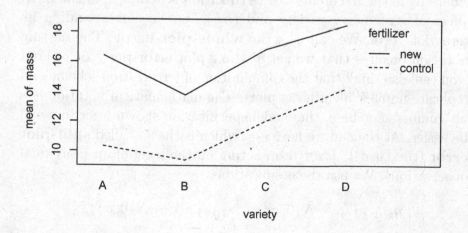

How can we model such data? To set up a correct model, we have to carefully study the randomization protocol that was applied. There were two randomizations involved here:

- fertilization schemes were randomized and applied to *plots* of land.
- strawberry varieties were randomized and applied to *subplots*.

Hence, an experimental unit for fertilizer is given by a *plot* of land, while for strawberry variety, the experimental unit is given by a *subplot*.

This design is an example of a **split-plot design**. Fertilization scheme is the so-called **whole-plot factor** and strawberry variety is the **split-plot factor**. A whole-plot is given by a plot of land and a split-plot by a subplot of land. For an illustration, see again Table 7.1.

As we have two different sizes of experimental units, we also need two error terms to model the corresponding experimental errors. We need one error term "acting" on the plot level and another one on the subplot level, which is the level of an individual observation. Let y_{ijk} be the observed mass of the kth replicate of a plot with fertilization scheme i and strawberry variety j. We use the model

$$Y_{ijk} = \mu + \alpha_i + \eta_{k(i)} + \beta_j + (\alpha\beta)_{ij} + \epsilon_{ijk}, \qquad (7.1)$$

where α_i is the fixed effect of fertilization scheme, β_j is the fixed effect of strawberry variety and $(\alpha\beta)_{ij}$ is the corresponding interaction term. We call $\eta_{k(i)}$ the **whole-plot error**. The nesting notation ensures that we get a whole-plot error per plot of land (you can also think that the combination of fertilization scheme and replicate number identifies a plot). The mathematical notation is a bit cumbersome here, the model specification shown later in R will be easier. At the end we have ϵ_{ijk} which is the so-called **split-plot error** (the "usual" error term acting on the level of an individual observation). We use the assumptions

$$\eta_{k(i)} \text{ i.i.d.} \sim N(0, \sigma_\eta^2), \quad \epsilon_{ijk} \text{ i.i.d.} \sim N(0, \sigma^2),$$

and independence between $\eta_{k(i)}$ and ϵ_{ijk}. The first part

$$\alpha_i + \eta_{k(i)}$$

of model formula (7.1) can be thought of as the "reaction" of an individual *plot* of land on the ith fertilization scheme. All plot specific properties are included in the whole-plot error $\eta_{k(i)}$. The fact that all subplots on the same plot share the same whole-plot error has the side effect that observations from the same plot are modeled as correlated data. The following part

$$\beta_j + (\alpha\beta)_{ij} + \epsilon_{ijk}$$

is the "reaction" of the *subplot* on the jth variety, including a potential interaction with the ith fertilization scheme. All subplot specific properties can now be found in the split-plot error ϵ_{ijk}.

If we only consider fertilization scheme, we do a completely randomized design here, with plots as experimental units. The first part of model formula (7.1) is actually the corresponding model equation of the corresponding one-way ANOVA. On the other hand, if we only consider variety, we could treat the plots as blocks and would have a randomized complete block design on this "level," including an interaction term; this is what we see in the second part of model formula (7.1).

To fit such a model in R, we use a mixed model approach. The whole-plot error, acting on plots, can easily be incorporated with (1 | plot). The split-plot error, acting on the subplot level, is automatically included, as it is on the level of individual observations. Hence, we end up with the following function call.

```
library(lmerTest)
fit.john <- lmer(mass ~ fertilizer * variety + (1 | plot),
                 data = john)
```

The F-tests can as usual be obtained with the function anova.

```
anova(fit.john)
```

```
## Type III Analysis of Variance Table with Satterthwaite's method
##                      Sum Sq Mean Sq NumDF DenDF F value
## fertilizer            137.4   137.4     1     6   68.24
## variety                96.4    32.1     3    18   15.96
## fertilizer:variety      4.2     1.4     3    18    0.69
##                      Pr(>F)
## fertilizer          0.00017
## variety             2.6e-05
## fertilizer:variety  0.56951
```

The interaction is not significant while the two main effects are. Let us have a closer look at the denominator degrees of freedom for a better understanding of the split-plot model. We observe that for the test of the main effect of fertilizer we only have six denominator degrees of freedom. Why? Because as described above, we basically performed a completely randomized design with eight experimental units (eight plots of land) and a treatment factor having two levels (control and new). Hence, the error has $8 - 1 - 1 = 6$ degrees of freedom. Alternatively, we could also argue that we should check whether the variation between different fertilization schemes is larger than the variation between plots getting the same fertilization scheme. The latter is given by the whole-plot error having (only!) $2 \cdot (4 - 1) = 6$ degrees of freedom, as it is nested in fertilizer.

In contrast, variety (and the interaction fertilizer:variety) are tested against the "usual error term": We have a total of 32 observations, leading to 31 degrees of freedom. We use 1 degree of freedom for fertilizer, 6 for the whole-plot error (see above), 3 for variety and another 3 for the interaction fertilizer:variety. Hence, the degrees of freedom that remain for the split-plot error are $31 - 1 - 6 - 3 - 3 = 18$. Another way of thinking is that we can interpret the experiment on the split-plot level as a randomized complete block design where we block on the different plots which uses 7 degrees of freedom. Both the main effect of variety and the interaction effect `fertilizer:variety` use another 3 degrees of freedom each such that we arrive at 18 degrees of freedom for the error term.

7.2 Properties of Split-Plot Designs

Why should we use split-plot designs? Typically, split-plot designs are suitable for situations where one of the factors can only be varied on a large scale. For example, fertilizer or irrigation on large plots of land. While "large" was literally large in the previous example, this is not always the case. Let us consider an example with a machine running under different settings using different source material. While it is easy to change the source material, it is much more tedious to change the machine settings. Or machine settings are hard to vary. Hence, we do not want to change them too often. We could think of an experimental design where we change the machine setting and keep using the same setting for different source materials. This means we are not completely randomizing machine setting and source material. This would be another example of a split-plot design where machine settings is the whole-plot factor and source material is the split-plot factor. Using this terminology, the factor which is hard to change will be the whole-plot factor.

The price that we pay for this "laziness" on the whole-plot level is less precision, or less power, for the corresponding main effect because we have much fewer observations on this level. We did *not* apply the whole-plot treatment very often; therefore, we cannot expect our results to be very precise. This can also be observed in the ANOVA table in Section 7.1. The denominator degrees of freedom of the main effect of `fertilizer` (the whole-plot factor) are only 6. Note that the main effect of the split-plot factor and the interaction between the split-plot and the whole-plot factor are not affected by this loss of efficiency. In fact, on the split-plot level, we are doing an efficient experiment as we block on the whole-plots (see also the explanation in Section 7.1).

Split-plot designs can be found quite often in practice. Identifying a split-plot design needs some experience. Often, a split-plot design was not done on purpose and hence the analysis does not take into account the special design structure and is therefore wrong. Typical signs for split-plot designs are:

- Some treatment factor is constant across multiple time-points, e.g., a whole week, while another changes at each time-point, e.g., each day.
- Some treatment factor is constant across multiple locations, e.g., a large plot of land, while another changes at each location, e.g., a subplot.
- When planning an experiment: Thoughts like, "It is easier if we do not change these settings too often …".

If we are not taking into account the special split-plot structure, the results on the whole-plot level will typically be overly optimistic, which means that p-values are too small, confidence interval are too narrow, etc. (see also the example later in Section 7.3). Again, there is no free lunch, this is the price that we pay for the "laziness." More information can for example be found in Goos et al. (2006).

Split-plot designs can of course arise in much more complex situations. We could, for example, extend the original design in the sense that we do a randomized complete block design or some factorial treatment structure on the whole-plot level. If we have more than two factors, we could also do a so-called split-split-plot design having one additional "layer," meaning that we would have three sizes of experimental units: whole plots, split plots and split-split plots. To specify the correct model, we simply have to inspect the randomization protocol. For every size of experimental unit we use a random effect as error term to model the corresponding experimental error. From a technical point of view, it is often helpful to define "helper variables" which define the corresponding experimental units. For example, we could have an extra variable PlotID which enumerates the different plots, as was the case with the original example in Section 7.1. A whole book about split-plot and related designs is Federer and King (2007). A discussion about efficiency considerations can for example be found in Jones and Nachtsheim (2009).

We conclude with an additional example.

7.3 A More Complex Example in Detail: Oat Varieties

To illustrate a more complex example, we consider the data set oats from the package MASS (Venables and Ripley, 2002). We actually already saw an aggregated version of this data set in Section 5.2.

```
data(oats, package = "MASS")
str(oats)
```

```
## 'data.frame':    72 obs. of  4 variables:
##  $ B: Factor w/ 6 levels "I","II","III",..: 1 1 1 1 1 1 ...
##  $ V: Factor w/ 3 levels "Golden.rain",..: 3 3 3 3 1 1 ...
##  $ N: Factor w/ 4 levels "0.0cwt","0.2cwt",..: 1 2 3 4 1 2..
##  $ Y: int  111 130 157 174 117 114 ...
```

The description in the help page (see ?oats) is, "The yield of oats from a split-plot field trial using three varieties and four levels of manurial treatment. The experiment was laid out in 6 blocks of 3 main plots, each split into 4 sub-plots. The varieties were applied to the main plots and the manurial treatments to the sub-plots." This means compared to Section 5.2, we now have an additional factor N which gives us information about the nitrogen (manurial) treatment. In Section 5.2 we simply used the average values across all nitrogen treatments as the response.

A visualization of the design for the first block can be found in Table 7.2. The whole-plot factor V (variety) is randomized and applied to plots (columns in Table 7.2), the split-plot factor N (nitrogen) is randomized and applied to subplots in each plot (cells within the same column in Table 7.2). See also Yates (1935) for a

TABLE 7.2: Potential layout of the design for block I. Plots are given by columns.

Block I		
Golden.rain	Marvellous	Victory
0.6 cwt	0.0 cwt	0.0 cwt
0.2 cwt	0.6 cwt	0.4 cwt
0.4 cwt	0.2 cwt	0.2 cwt
0.0 cwt	0.4 cwt	0.6 cwt

more detailed description of the actual layout (which was in fact a 2-by-2 layout for the subplots).

Interaction plots for all blocks can be easily produced with the package ggplot2.

```
library(ggplot2)
ggplot(aes(x = N, y = Y, group = V, lty = V), data = oats) +
  geom_line() + facet_wrap(~ B) + theme_bw()
```

In Figure 7.1 we can observe that blocks are different (this is why we use them), there is no clear effect of variety (v), but there seems to be a more or less linear effect of nitrogen (N).

What model can we set up here? Let us again have a closer look at the randomization scheme that was applied. With respect to variety we have a randomized complete block design. This is the whole-plot level, and we need to include the corresponding whole-plot error (on each plot).

Here, a plot can be identified by the combination of the levels of B and V. Hence, the whole-plot error can be written as (1 | B:V). This leads to the following call of lmer.

```
fit.oats <- lmer(Y ~ B + V * N + (1 | B:V), data = oats)
```

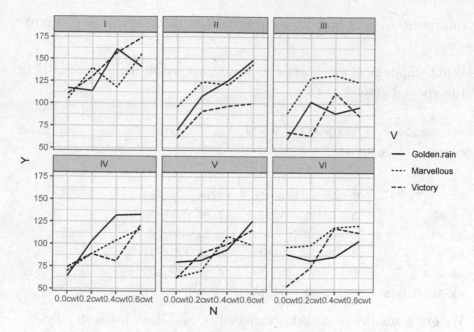

FIGURE 7.1: Interaction plots for each block of the oats data set. Blocks are labelled with Roman numerals.

```
anova(fit.oats)
```

```
## Type III Analysis of Variance Table with Satterthwaite's method
##        Sum Sq Mean Sq NumDF DenDF F value  Pr(>F)
## B        4675     935     5    10    5.28   0.012
## V         526     263     2    10    1.49   0.272
## N       20021    6674     3    45   37.69 2.5e-12
## V:N       322      54     6    45    0.30   0.932
```

Again, we see that we only get 10 degrees of freedom for the test of variety (v). In fact, the result is identical with what we got in Section 5.2! This again illustrates the way of thinking when analyzing split-plot designs: Think of averaging away the nitrogen information by taking the sample mean on each plot. With this *reduced* set of values, perform a classical analysis of a randomized complete block design (this is exactly what we did in Section 5.2). Regarding the split-plot factor: nitrogen (N) is significant, but the interaction V:N is not. This confirms the interpretation of Figure 7.1.

What happens if we choose the wrong approach using aov (without the special error term)?

```
fit.oats.aov <- aov(Y ~ B + V * N, data = oats)
summary(fit.oats.aov)
```

```
##              Df Sum Sq Mean Sq F value  Pr(>F)
## B             5  15875    3175   12.49 4.1e-08
## V             2   1786     893    3.51   0.037
## N             3  20020    6673   26.25 1.1e-10
## V:N           6    322      54    0.21   0.972
## Residuals    55  13982     254
```

We get a smaller p-value for variety (V), and if we use a significance level of 5%, variety would now be significant! The reason behind this is that aov thinks that we randomized and applied the different varieties on individual subplots. Hence, the corresponding error estimate is too small and the results are overly optimistic. The model thinks we used 72 experimental units (subplots), whereas in practice we only used 18 (plots) for variety.

8

Incomplete Block Designs

8.1 Introduction

In this chapter we are going to learn more about block designs. The block designs in Chapter 5 were *complete*, meaning that every block contained *all* treatments. In practice, this is not always possible. For example, the actual physical size of a block might be too small. There are also situations where it is not advisable to have too many treatments in each block. If we do a food tasting experiment, we typically want to restrict the number of different recipes (treatments) we want an individual rater to taste. Otherwise, the response will get unreliable, for example, because the raters might get bored after some time.

If we cannot "afford" to do a complete design, we get a so-called **incomplete block design (IBD)**. We have to decide what *subset* of treatments we use in an individual block. If we do not make a wise decision, it can be that the design is flawed in the sense that certain quantities are not estimable anymore (e.g., some treatment differences). This would be a worst case scenario.

Consider the design in Table 8.1 with treatments A to F and blocks 1 to 6 (each *column* corresponds to a block). For example, on block 5 we apply the two treatments D and F. Think for example of treatments as different recipes and block as different raters.

TABLE 8.1: Example of an incomplete block design. Blocks are denoted by different columns, treatments by the letters A to F.

1	2	3	4	5	6
A	A	B	D	D	E
B	C	C	E	F	F

First, we create a fictional data set having the same structure as in Table 8.1.

```
set.seed(25) ## for reproducibility
dat <- data.frame(block = factor(rep(1:6, times = 2)),
                  treat = factor(c("A", "A", "B", "D", "D", "E",
                                   "B", "C", "C", "E", "F", "F")),
                  ## block eff. + error term, no signal wrt. treat
                  y = round(rep(rnorm(6), times = 2) +
                            rnorm(12), 1))
str(dat)
```

```
## 'data.frame':    12 obs. of  3 variables:
##  $ block: Factor w/ 6 levels "1","2","3","4",..: 1 2 3 4 5..
##  $ treat: Factor w/ 6 levels "A","B","C","D",..: 1 1 2 4 4..
##  $ y    : num  1.6 1.7 0.9 0.7 -0.8 3.4 ...
```

We can get an overview of the used combinations with xtabs.

```
xtabs(~ treat + block, data = dat)
```

```
##      block
## treat 1 2 3 4 5 6
##     A 1 1 0 0 0 0
##     B 1 0 1 0 0 0
##     C 0 1 1 0 0 0
##     D 0 0 0 1 1 0
##     E 0 0 0 1 0 1
##     F 0 0 0 0 1 1
```

What happens if we fit our standard main effects model here?

```
options(contrasts = c("contr.treatment", "contr.poly"))
fit.ibd <- aov(y ~ block + treat, data = dat)
dummy.coef(fit.ibd)
```

```
## Full coefficients are
##
## (Intercept):    1.1
## block:          1    2    3    4    5    6
##                 0.0  1.1  0.5 -2.9 -4.0  1.1
## treat:          A    B    C    D    E    F
##                 0.0 -1.2 -1.8  2.3  1.0  0.0
```

At first sight, it looks like we were able to fit the specified model. However, it is strange that *two* of the treatment coefficients are set to zero and not only *one* as usual when using contr.treatment.

What happens if we want to investigate a certain contrast, say treatment A vs. F?

```
library(multcomp)
summary(glht(fit.ibd, linfct = mcp(treat = c(1, 0, 0, 0, 0, -1))))
```

We get an error message (not shown) which says that some linear functions are not estimable. In fact, we *cannot* estimate the difference between the treatments A and F with this data set, at least not with a fixed effects model. The reason lies in the structure of the incomplete design. It is an example of a so-called **disconnected design**. We have the treatment set $\{A, B, C\}$ in the first three blocks and $\{D, E, F\}$ in the last three blocks (check the output of xtabs above). As the two sets are disjoint, we have no idea about any differences involving treatments from the two sets, because we never see a pair together in the same block. This is also causing the strange output in dummy.coef above. Because of this dependency between treatment and block, we cannot distinguish block and treatment effects anymore; the effects are confounded. Here, R automatically drops one of the treatment levels from the model, which might be very dangerous if you are not carefully inspecting the output.

Hence, performing such a disconnected design was obviously not a good idea from a design of experiment point of view, as we cannot estimate any differences involving treatments from the two disjoint sets. Still, situations like these happen in practice: "Let's do treatments A, B on the first few days and then switch to C, D because it is easier ...". If day is a block factor, this design will be disconnected.

Intuitively, we should have a good "mix" of treatments in each block.

8.2 Balanced Incomplete Block Designs

We could simply randomize subsets of treatments to different blocks. This would work well if we have enough blocks. However, if we only have a small number of blocks, there would be the risk that we end up with a disconnected design.

On the other hand, we can also try to fulfill some optimality criterion. One candidate would be the criterion that we can estimate all treatment differences with the **same precision**. For example, this would mean that all confidence intervals for $\alpha_i - \alpha_j$ have the same width (for *any* pair i, j, where α_i is the ith treatment effect). The design which fulfills this criterion is the so-called balanced incomplete block design.

A **balanced incomplete block design (BIBD)** is an *incomplete* block design where all *pairs* of treatments occur together in the *same* block equally often; we denote this number by λ.

Using an example where blocks are given by raters and treatments by different chocolate chip cookie brands, we introduce the following notation, similar to Oehlert (2000):

- g: number of treatments ("number of different chocolate chip cookie brands")
- b: number of blocks ("number of raters")
- k: block size; number of experimental units per block ($k < g$) ("number of cookie brands each rater gets to taste")
- r: number of replicates per treatment ("how often do we observe a certain cookie brand across all raters?")
- N: total number of experimental units

By definition, we have $N = b \cdot k = g \cdot r$. How can we construct a BIBD? For every setting $k < g$ we can find a BIBD by taking *all* possible subsets, where we have $\binom{g}{k}$ (binomial coefficient, implemented in the function choose in R). The corresponding design is called an **unreduced balanced incomplete block design**. For example, if we have $g = 6$ treatments and $k = 3$ experimental units per block, we get $\binom{6}{3} = 20$ blocks. In R, we can easily get this with the function combn.

```
combn(x = 6, m = 3)
```

```
##         [,1] [,2] [,3] [,4] [,5] [,6] [,7] [,8] [,9] [,10]
## [1,]     1    1    1    1    1    1    1    1    1    1
## [2,]     2    2    2    2    3    3    3    4    4    5
## [3,]     3    4    5    6    4    5    6    5    6    6
##         [,11] [,12] [,13] [,14] [,15] [,16] [,17] [,18] [,19]
## [1,]     2    2    2    2    2    2    3    3    3
## [2,]     3    3    3    4    4    5    4    4    5
## [3,]     4    5    6    5    6    6    5    6    6
##         [,20]
## [1,]     4
## [2,]     5
## [3,]     6
```

In the matrix above, every column defines a block. For example, block 10 would get treatments 1, 5 and 6. Typically, we would randomize the order or the placement of the treatments within a block, as we should always do.

In practice, we cannot always afford to do an unreduced BIBD, as the required number of blocks might be too large, for example, if we do not have enough raters. Whether a BIBD exists for a certain desired setting of number of blocks b, block size k and number of treatments g, is a combinatorial problem. A necessary, but not sufficient, condition for a BIBD to exist is

$$\frac{r \cdot (k-1)}{g-1} = \lambda, \tag{8.1}$$

where λ is the number of times two treatments occur together in the same block (hence, an integer).

What is the intuition behind Equation (8.1)? A treatment appears in a total of r different blocks. In each of these blocks, there are $k - 1$ available other experimental units, leading to a total of $r \cdot (k - 1)$ available experimental units. There are $g - 1$ other available treatments that we must divide among them. Hence, it must hold that $r \cdot (k-1) = \lambda \cdot (g-1)$; otherwise, it is impossible to construct a BIBD. As already mentioned above, this condition is necessary, but not sufficient. This means that even if the condition is fulfilled, it might be impossible to construct a BIBD.

In R, package ibd (Mandal, 2019) provides some functionality to find (B)IBDs. For example, function bibd searches for a balanced incomplete block design using sophisticated numerical optimization techniques.

```
library(ibd)
des.bibd <- bibd(v = 6, b = 10, r = 5, k = 3, lambda = 2)
## arguments of function bibd are:
## - v: number of treatments
## - b: number of blocks
## - r: number of replicates (across all blocks)
## - k: number of experimental units per block
## - lambda: lambda

des.bibd$design ## here, blocks are given by *rows*
```

```
##              [,1] [,2] [,3]
## Block-1        2    3    6
## Block-2        2    5    6
## Block-3        4    5    6
## Block-4        3    4    5
## Block-5        1    4    6
## Block-6        1    3    6
## Block-7        1    3    5
## Block-8        1    2    5
## Block-9        2    3    4
## Block-10       1    2    4
```

In des.bibd$design, each *row* corresponds to a block. For a setting where no BIBD is possible, we can try to find a design that is nearly balanced using the function ibd (function bibd would return "parameters do not satisfy necessary conditions" in such a situation).

```
des.ibd <- ibd(v = 6, b = 9, k = 3)
## - v: number of treatments
## - b: number of blocks
## - k: number of units per block
des.ibd$design
```

```
##              [,1] [,2] [,3]
## Block-1        1    5    6
## Block-2        1    4    5
## Block-3        3    5    6
## Block-4        3    4    6
## Block-5        1    3    4
## Block-6        1    2    6
## Block-7        2    4    6
## Block-8        2    4    5
## Block-9        2    3    5
```

We can find the so-called concurrence matrix of the generated design in des.ibd$conc.mat. It contains information about how many times any pair of treatments appears together in the same block.

```
des.ibd$conc.mat ## check for (un)balancedness
```

```
##        [,1] [,2] [,3] [,4] [,5] [,6]
## [1,]    4    1    1    2    2    2
## [2,]    1    4    1    2    2    2
## [3,]    1    1    4    2    2    2
## [4,]    2    2    2    5    2    2
## [5,]    2    2    2    2    5    2
## [6,]    2    2    2    2    2    5
```

On the diagonal, we can read off how often every treatment appears across all blocks. In the off-diagonal elements, we have the corresponding information about treatment pairs. For example, treatment 1 and 4 appear together twice in the same block. On the other hand, treatment 1 and 3 appear together only once.

For a given design, we would then randomize the different treatment subsets, i.e., the different rows in des.ibd$design, to the actual blocks, e.g., raters. Within each block, we would randomize the corresponding treatments to the experimental units, e.g., time-slots, and last but not least we would globally randomize the treatment numbers to the actual treatments, e.g., cookie brands.

8.3 Analysis of Incomplete Block Designs

8.3.1 Example: Taste Data

The analysis of an incomplete block design is "as usual." We use a fixed block factor and a treatment factor leading to

$$Y_{ij} = \mu + \alpha_i + \beta_j + \epsilon_{ij}, \tag{8.2}$$

where the α_i's are the treatment effects and the β_j's are the block effects with the usual side constraints. Because we do not observe all the block and treatment combinations equally often (some are simply missing), we are faced with an *unbalanced* design. We typically use sum of squares for treatment effects that are adjusted for block effects.

Let us consider the data set taste from package daewr (Lawson and Krennrich, 2021) as an example. Twelve different panelists each rated two out of four different recipes. We first load the data and try to get an overview.

```
data("taste", package = "daewr")
str(taste)
```

```
## 'data.frame':    24 obs. of  3 variables:
##  $ panelist: Factor w/ 12 levels "1","2","3","4",..: 1 2 3..
##  $ recipe  : Factor w/ 4 levels "A","B","C","D": 1 1 1 2 2..
##  $ score   : num  5 7 5 6 6 8 ...
```

```
(tab <- xtabs(~ recipe + panelist, data = taste))
```

```
##         panelist
## recipe 1 2 3 4 5 6 7 8 9 10 11 12
##      A 1 1 1 0 0 0 1 1 1  0  0  0
##      B 1 0 0 1 1 0 1 0 0  1  1  0
##      C 0 1 0 1 0 1 0 1 0  1  0  1
##      D 0 0 1 0 1 1 0 0 1  0  1  1
```

The output of xtabs tells us that we have an *incomplete* design here. It is actually a balanced incomplete block design. We can test this with the function isGYD in package crossdes (Sailer, 2013). For that reason, we first have to transform the design information to the desired form such that we have a matrix with 12 rows (panelists) and two columns containing the two treatments. This can be done with the following rather technical code where we extract the corresponding treatments column-wise.

```
library(crossdes)
(d <- t(apply(tab, 2, function(x) (1:4)[x != 0])))
```

```
##
## panelist [,1] [,2]
##        1     1    2
##        2     1    3
##        3     1    4
##        4     2    3
##        5     2    4
##        6     3    4
##        7     1    2
##        8     1    3
##        9     1    4
##       10     2    3
##       11     2    4
##       12     3    4
```

```
isGYD(d)
```

```
##
## [1] The design is a balanced incomplete block design w.r.t. rows.
```

We use the model in Equation (8.2). As it is an unbalanced data set, we use drop1 such that we get the sum of squares of recipe adjusted for panelist. We could also use summary here because recipe appears last in the model formula.

```
fit.taste <- aov(score ~ panelist + recipe, data = taste)
drop1(fit.taste, test = "F")
```

```
## Single term deletions
##
## Model:
## score ~ panelist + recipe
##          Df Sum of Sq    RSS    AIC F value Pr(>F)
## <none>                  6.87   0.00
## panelist 11      7.46 14.33  -4.37    0.89  0.581
## recipe    3      9.13 16.00  14.27    3.98  0.046
```

The *F*-test of recipe is significant. To illustrate, we have a look at all pairwise comparisons without any multiple testing correction.

```
library(multcomp)
contr <- glht(fit.taste, linfct = mcp(recipe = "Tukey"))
summary(contr, test = adjusted("none"))
```

```
##
## Simultaneous Tests for General Linear Hypotheses
##
## Multiple Comparisons of Means: Tukey Contrasts
## ...
## Linear Hypotheses:
##              Estimate Std. Error t value Pr(>|t|)
## B - A == 0     0.750     0.618     1.21    0.256
## C - A == 0     1.375     0.618     2.22    0.053
## D - A == 0    -0.625     0.618    -1.01    0.338
## C - B == 0     0.625     0.618     1.01    0.338
## D - B == 0    -1.375     0.618    -2.22    0.053
## D - C == 0    -2.000     0.618    -3.24    0.010
## (Adjusted p values reported -- none method)
```

We are not inspecting the p-values here, as it is a multiple testing problem. However, what we can observe is the fact that *all* differences share the *same* standard error (remember the motivation for a BIBD?). This is a feature of this special design!

8.3.2 Intra- and Inter-block Analysis

Up until now, we estimated treatment effects by adjusting for block effects. This means that whatever is special to a block is fully allocated to the block effect and does *not* affect the treatment effect. Basically, the estimate of the treatment effect is based on the "leftovers." This is also known as an **intra-block analysis**.

On the other hand, if we treat the block factor as a random effect, the mean of the values of a block implicitly also contain information about the treatment effects. An analysis which is based on this information is known as an **inter-block analysis**. This leads to another estimate of the treatment effects. Both approaches can be combined. In practice, this can be achieved by using lmer and setting the block factor as a random effect. Typically, there is only little benefit; see also the discussion in section 14.1.2 of Oehlert (2000).

However, there are also situations where the effect can be dramatic. If we consider the introductory example with the disconnected design, some magic happens:

```
library(lmerTest)
fit.ibd2 <- lmer(y ~ treat + (1 | block), data = dat)
summary(glht(fit.ibd2, linfct = mcp(treat = c(1, 0, 0, 0, 0, -1))))
```

```
##
##   Simultaneous Tests for General Linear Hypotheses
##
## Multiple Comparisons of Means: User-defined Contrasts
## ...
## Linear Hypotheses:
##        Estimate Std. Error z value Pr(>|z|)
## 1 == 0     2.39       1.77    1.35     0.18
## (Adjusted p values reported -- single-step method)
```

Here it is: an estimate of the difference between A and F! However, this is standing on very thin ice and you should not rely on this trick in practice!

8.4 Outlook

A balanced incomplete block design is of course only a very special case of an incomplete block design. A more general version is a so-called **partially balanced incomplete block design (PBID)** where some treatment pairs occur together more often than other pairs (defined through so-called associate classes). Hence, not all treatment differences are estimated with the same precision. This is useful if more precision is needed for comparisons with a control treatment than for comparisons between the other treatments. See for example Kuehl (2000) and Oehlert (2000) for more details.

With two block factors we can also create block designs which are incomplete in one or both "directions," so-called row-column designs, see also Section 5.4. An example would be a design where the rows form a complete block design but the columns an incomplete block design. Again, more details can be found in Kuehl (2000) and Oehlert (2000).

8.5   Concluding Remarks

We have now learned about some of the more important experimental designs and the corresponding statistical analyses. The last few chapters were mostly in the form of short introductions (as it says in the book title) and more information can be found in the references mentioned in the corresponding chapters. A good place to start looking for more information about R packages are the CRAN task views[1] which provide a nice overview of packages that are suitable for special topics. There is a task view "Design of Experiments (DoE) & Analysis of Experimental Data"[2] which contains some of the packages that we already used in this book, but of course also many more.

---

[1]https://cran.r-project.org/web/views/
[2]https://cran.r-project.org/web/views/ExperimentalDesign.html

Bibliography

Anderson, M. J. (2001). Permutation tests for univariate or multivariate analysis of variance and regression. *Canadian Journal of Fisheries and Aquatic Sciences*, 58(3):626–639.

Bates, D., Mächler, M., Bolker, B., and Walker, S. (2015). Fitting linear mixed-effects models using lme4. *Journal of Statistical Software*, 67(1):1–48.

Benjamini, Y., Heller, R., and Yekutieli, D. (2009). Selective inference in complex research. *Philosophical Transactions of the Royal Society A: Mathematical, Physical and Engineering Sciences*, 367(1906):4255–4271.

Box, G. E. P. (1953). Non-normality and tests on variances. *Biometrika*, 40(3/4):318–335.

Box, G. E. P. and Cox, D. R. (1964). An analysis of transformations. *Journal of the Royal Statistical Society. Series B (Methodological)*, 26(2):211–252.

Box, G. E. P., Hunter, W. G., and Hunter, J. S. (1978). *Statistics for experimenters: an introduction to design, data analysis, and model building*. Wiley series in probability and mathematical statistics. Wiley, New York, NY.

Bretz, F., Hothorn, T., and Westfall, P. H. (2011). *Multiple comparisons using R*. CRC Press, Boca Raton, FL.

Brooks, M. E., Kristensen, K., van Benthem, K. J., Magnusson, A., Berg, C. W., Nielsen, A., Skaug, H. J., Maechler, M., and Bolker, B. M. (2017). glmmTMB balances speed and flexibility among packages for zero-inflated generalized linear mixed modeling. *The R Journal*, 9(2):378–400.

Burdick, R. K. and Graybill, F. A. (1992). *Confidence intervals on variance components.* Number 127 in Statistics, textbooks and monographs. M. Dekker, New York.

Dalgaard, P. (2008). *Introductory statistics with R.* Statistics and computing. Springer, New York, NY, 2nd edition.

de Mendiburu, F. (2020). *agricolae: Statistical Procedures for Agricultural Research.* R package version 1.3-3.

Dobson, A. (1983). *Introduction to Statistical Modelling.* Springer, New York, NY.

Dunnett, C. W. (1955). A multiple comparison procedure for comparing several treatments with a control. *Journal of the American Statistical Association*, 50(272):1096–1121.

Edgington, E. and Onghena, P. (2007). *Randomization Tests.* Statistics: A Series of Textbooks and Monographs. CRC Press, Boca Raton, FL.

Faraway, J. J. (2005). *Linear models with R.* Texts in statistical science. Chapman & Hall/CRC, Boca Raton, FL.

Federer, W. T. and King, F. (2007). *Variations on split plot and split block experiment designs.* Wiley series in probability and statistics. Wiley-Interscience, Hoboken, NJ.

Fitzmaurice, G. M., Laird, N. M., and Ware, J. H. (2011). *Applied Longitudinal Analysis.* Wiley Series in Probability and Statistics. John Wiley & Sons, Inc., Hoboken, NJ.

Fox, J. and Weisberg, S. (2019). *An R Companion to Applied Regression.* Sage, Thousand Oaks, CA, third edition.

Goos, P., Langhans, I., and Vandebroek, M. (2006). Practical inference from industrial split-plot designs. *Journal of Quality Technology*, 38(2):162–179.

Green, P. and MacLeod, C. J. (2016). simr: an R package for power analysis of generalised linear mixed models by simulation. *Methods in Ecology and Evolution*, 7(4):493–498.

Hayter, A. J. (1984). A proof of the conjecture that the Tukey-Kramer multiple comparisons procedure is conservative. *The Annals of Statistics*, 12(1):61–75.

Henderson, C. R. (1963). Selection index and expected genetic advance. In Hanson, W. and Robinson, H., editors, *Statistical Genetics and Plant Breeding*, number 982, pages 141–163. National Academy of Sciences-National Research Council, Washington, D.C.

Holm, S. (1979). A simple sequentially rejective multiple test procedure. *Scandinavian Journal of Statistics*, 6(2):65–70.

Hothorn, T., Bretz, F., and Westfall, P. (2008). Simultaneous inference in general parametric models. *Biometrical Journal*, 50(3):346–363.

Hsu, J. (1996). *Multiple Comparisons: Theory and Methods*. Guilford School Practitioner. Chapman and Hall/CRC Press, Boca Raton, FL.

Huitema, B. E. (2011). *The Analysis of Covariance and Alternatives: Statistical Methods for Experiments, Quasi-Experiments, and Single-Case Studies*. Wiley Series in Probability and Statistics. John Wiley & Sons, Inc., Hoboken, NJ.

Jiang, J. and Nguyen, T. (2021). *Linear and Generalized Linear Mixed Models and Their Applications*. Springer Series in Statistics. Springer, New York, NY.

Jones, B. and Kenward, M. (2014). *Design and analysis of crossover trials*. Number 138 in Chapman & Hall / CRC monographs on statistics and applied probability. CRC Press/Taylor & Francis, Boca Raton, FL, third edition.

Jones, B. and Nachtsheim, C. (2009). Split-plot designs: What, why, and how. *Journal of Quality Technology*, 41(4):340–361.

Keane, B. (1998). The family circus. Cowles Syndicate. Comic strip.

Kramer, C. Y. (1956). Extension of multiple range tests to group means with unequal numbers of replications. *Biometrics*, 12(3):307.

Kuehl, R. (2000). *Design of Experiments: Statistical Principles of Research Design and Analysis*. Statistics Series. Duxbury/Thomson Learning.

Kuznetsova, A., Brockhoff, P. B., and Christensen, R. H. B. (2017). lmerTest package: Tests in linear mixed effects models. *Journal of Statistical Software*, 82(13):1–26.

Lakens, D. and Caldwell, A. (2021). Simulation-based power analysis for factorial analysis of variance designs. *Advances in Methods and Practices in Psychological Science*, 4(1):1–14.

Lawson, J. (2014). *Design and Analysis of Experiments with R*. Chapman & Hall/CRC Texts in Statistical Science. CRC Press/Taylor & Francis, Boca Raton, FL.

Lawson, J. and Krennrich, G. (2021). *daewr: Design and Analysis of Experiments with R*. R package version 1.2-7.

Lenth, R. V. (2020). *emmeans: Estimated Marginal Means, aka Least-Squares Means*. R package version 1.5.3.

Loy, A. and Hofmann, H. (2014). HLMdiag: A suite of diagnostics for hierarchical linear models in R. *Journal of Statistical Software*, 56(5):1–28.

Loy, A. and Hofmann, H. (2015). Are you normal? The problem of confounded residual structures in hierarchical linear models. *Journal of Computational and Graphical Statistics*, 24(4):1191–1209.

Loy, A., Hofmann, H., and Cook, D. (2017). Model choice and diagnostics for linear mixed-effects models using statistics on street corners. *Journal of Computational and Graphical Statistics*, 26(3):478–492.

Mandal, B. N. (2019). *ibd: Incomplete Block Designs*. R package version 1.5.

Montgomery, D. (2019). *Design and Analysis of Experiments.* Wiley, Hoboken, NJ.

Oehlert, G. (2000). *A First Course in Design and Analysis of Experiments.* W.H. Freeman, New York, NY.

Pearl, J. and Mackenzie, D. (2018). *The Book of Why: The New Science of Cause and Effect.* Basic Books, Inc., New York, NY, 1st edition.

Pinheiro, J. and Bates, D. (2009). *Mixed-Effects Models in S and S-PLUS.* Statistics and Computing. Springer, New York, NY.

Pinheiro, J., Bates, D., DebRoy, S., Sarkar, D., and R Core Team (2021). *nlme: Linear and Nonlinear Mixed Effects Models.* R package version 3.1-152.

R Core Team (2021). *R: A Language and Environment for Statistical Computing.* R Foundation for Statistical Computing, Vienna, Austria.

Rice, J. (2007). *Mathematical statistics and data analysis.* Advanced series. Cengage Learning, Boston, MA.

Robinson, G. K. (1991). That BLUP is a good thing: The estimation of random effects. *Statistical Science*, 6(1):15–32.

Sailer, M. O. (2013). *crossdes: Construction of Crossover Designs.* R package version 1.1-1.

Santos Nobre, J. and da Motta Singer, J. (2007). Residual analysis for linear mixed models. *Biometrical Journal*, 49(6):863–875.

Scheffé, H. (1959). *The Analysis of Variance.* John Wiley & Sons, New York, NY.

Scheipl, F., Greven, S., and Kuechenhoff, H. (2008). Size and power of tests for a zero random effect variance or polynomial regression in additive and linear mixed models. *Computational Statistics & Data Analysis*, 52(7):3283–3299.

Signorell, A. et al. (2021). *DescTools: Tools for Descriptive Statistics.* R package version 0.99.43.

Speed, M., Hocking, R., and Hackney, P. (1978). Methods of analysis of linear models with unbalanced data. *Journal of the American Statistical Association*, 73(361):105–112.

Stolley, P. D. (1991). When genius errs: R. A. Fisher and the lung cancer controversy. *American Journal of Epidemiology*, 133(5):416–425.

Tukey, J. W. (1949a). Comparing individual means in the analysis of variance. *Biometrics*, 5(2):99–114.

Tukey, J. W. (1949b). One degree of freedom for non-additivity. *Biometrics*, 5(3):232–242.

Venables, W. N. and Ripley, B. D. (2002). *Modern Applied Statistics with S*. Springer, New York, NY, fourth edition. ISBN 0-387-95457-0.

Wickham, H. (2016). *ggplot2: Elegant Graphics for Data Analysis*. Springer, New York, NY.

Xie, Y. (2015). *Dynamic Documents with R and knitr*. Chapman & Hall/CRC Press, Boca Raton, FL, 2nd edition. ISBN 978-1498716963.

Xie, Y. (2021). *bookdown: Authoring Books and Technical Documents with R Markdown*. R package version 0.22.

Yates, F. (1935). Complex experiments. *Supplement to the Journal of the Royal Statistical Society*, 2(2):181–247.

Index

Printed in the United States
by Baker & Taylor Publisher Services